Problem Books in Mathematics

Edited by P. Winkler

For other titles in this series, go to
http://www.springer.com/series/714

Asuman G. Aksoy
Mohamed A. Khamsi

A Problem Book in Real Analysis

Asuman G. Aksoy
Department of Mathematics
Claremont McKenna College
Claremont, CA 91711
USA
aaksoy@cmc.edu

Mohamed A. Khamsi
Department of Mathematical Sciences
University of Texas at El Paso
El Paso, TX 79968
USA
mohamed@utep.edu

Series Editor:
Peter Winkler
Department of Mathematics
Dartmouth College
Hanover, NH 03755
USA
peter.winkler@dartmouth.edu

ISSN 0941-3502
ISBN 978-1-4419-1295-4   e-ISBN 978-1-4419-1296-1
DOI 10.1007/978-1-4419-1296-1
Springer New York Dordrecht Heidelberg London

Library of Congress Control Number: 2009939759

Mathematics Subject Classification (2000): 00A07

© Springer Science+Business Media, LLC 2010
All rights reserved. This work may not be translated or copied in whole or in part without the written permission of the publisher (Springer Science+Business Media, LLC, 233 Spring Street, New York, NY 10013, USA), except for brief excerpts in connection with reviews or scholarly analysis. Use in connection with any form of information storage and retrieval, electronic adaptation, computer software, or by similar or dissimilar methodology now known or hereafter developed is forbidden.
The use in this publication of trade names, trademarks, service marks, and similar terms, even if they are not identified as such, is not to be taken as an expression of opinion as to whether or not they are subject to proprietary rights.

Printed on acid-free paper

Springer is part of Springer Science+Business Media (www.springer.com)

Dedicated to Ercüment G. Aksoy and Anny Morrobel-Sosa

Contents

Preface	ix
1 Elementary Logic and Set Theory	**1**
Solutions	9
2 Real Numbers	**21**
Solutions	27
3 Sequences	**41**
Solutions	47
4 Limits of Functions	**63**
Solutions	68
5 Continuity	**77**
Solutions	84
6 Differentiability	**97**
Solutions	105
7 Integration	**127**
Solutions	136
8 Series	**159**
Solutions	166
9 Metric Spaces	**181**
Solutions	186
10 Fundamentals of Topology	**197**
Solutions	206
11 Sequences and Series of Functions	**223**
Solutions	231
Bibliography	**249**
Index	**251**

Preface

Education is an admirable thing, but it is well to remember from time to time that nothing worth knowing can be taught.

Oscar Wilde, "The Critic as Artist," 1890.

Analysis is a profound subject; it is neither easy to understand nor summarize. However, Real Analysis can be discovered by solving problems. This book aims to give independent students the opportunity to discover Real Analysis by themselves through problem solving.

The depth and complexity of the theory of Analysis can be appreciated by taking a glimpse at its developmental history. Although Analysis was conceived in the 17th century during the Scientific Revolution, it has taken nearly two hundred years to establish its theoretical basis. Kepler, Galileo, Descartes, Fermat, Newton and Leibniz were among those who contributed to its genesis. Deep conceptual changes in Analysis were brought about in the 19th century by Cauchy and Weierstrass. Furthermore, modern concepts such as open and closed sets were introduced in the 1900s.

Today nearly every undergraduate mathematics program requires at least one semester of Real Analysis. Often, students consider this course to be the most challenging or even intimidating of all their mathematics major requirements. The primary goal of this book is to alleviate those concerns by systematically solving the problems related to the core concepts of most analysis courses. In doing so, we hope that learning analysis becomes less taxing and thereby more satisfying.

The wide variety of exercises presented in this book range from the computational to the more conceptual and vary in difficulty. They cover the following subjects: Set Theory, Real Numbers, Sequences, Limits of Functions, Continuity, Differentiability, Integration, Series, Metric Spaces, Sequences and Series of Functions and Fundamentals of Topology. Prerequisites for accessing this book are a robust understanding of Calculus and Linear Algebra. While we define the concepts and cite theorems used in each chapter, it is best to use this book alongside standard analysis books such as: *Principles of Mathematical Analysis* by W. Rudin, *Understanding Analysis* by S. Abbott, *Elementary Classical Analysis* by J. E. Marsden and M. J. Hoffman, and *Elements of Real Analysis* by D. A. Sprecher. A list of analysis texts is provided at the end of the book.

Although *A Problem Book in Real Analysis* is intended mainly for undergraduate mathematics students, it can also be used by teachers to enhance their lectures or as an aid in preparing exams. The proper way to use this book is for students to first attempt to solve its problems without looking at solutions. Furthermore, students should try to produce solutions which are different from those presented in this book. It is through the search for a solution that one learns most mathematics.

Knowledge accumulated from many analysis books we have studied in the past has surely influenced the solutions we have given here. Giving proper credit to all the contributors is a difficult task that we have not undertaken; however, they are all appreciated. We also thank Claremont students Aaron J. Arvey, Vincent E. Selhorst-Jones and Martijn van Schaardenburg for their help with LaTeX. The source for the photographs and quotes given at the beginning of each chapter in this book are from the archive at http://www-history.mcs.st-andrews.ac.uk/

Perhaps Oscar Wilde is correct in saying "nothing worth knowing can be taught." Regardless, teachers can show that there are paths to knowledge. This book is intended to reveal such a path to understanding Real Analysis. *A Problem Book in Real Analysis* is not simply a collection of problems; it intends to stimulate its readers to independent thought in discovering Analysis.

Asuman Güven Aksoy
Mohamed Amine Khamsi
May 2009

Chapter 1

Elementary Logic and Set Theory

Reserve your right to think, for even to think wrongly is better than not to think at all.

Hypatia of Alexandria (370–415)

- If x belongs to a class A, we write $x \in A$ and read as "x is an element of A." Otherwise, we write $x \notin A$.

- If A and B are sets, then $A \subseteq B$ ("A is a subset of B" or "A is contained in B") means that each element of A is also an element of B. Sometimes we write $B \supseteq A$ ("B contains A") instead of $A \subseteq B$.

- We say two sets A and B are *equal*, written $A = B$, if $A \subseteq B$ and $B \subseteq A$.

- Any statement S has a *negation* $\sim S$ ("not S") defined by

  $\sim S$ is true if S is false   and $\sim S$ is false if S is true.

- Let $P(x)$ denote a *property* P of the object x. We write $\exists$ for the quantifier "*there exists*." The expression

  $$\exists\, x \in X : P(x)$$

  means that "there exists (at least) one object x in the class X which has the property P." The symbol $\exists$ is called the *existential quantifier*.

- We use the symbol $\forall$ for the quantifier *"for all."* The expression

$$\forall x \in X \; : \; P(x)$$

has the meaning "for each object x in the class X, x has property P." The symbol $\forall$ is called the *universal quantifier* (or sometimes the *general quantifier*).

- We use the symbol := to mean *"is defined by."* We take $x := y$ to mean that the object or symbol x is defined by the expression y.

- Note that for negation of a statement we have:

  (i) $\sim\sim A := \sim(\sim A) = A$
  (ii) $\sim(A \text{ and } B) = (\sim A) \text{ or } (\sim B)$
  (iii) $\sim(A \text{ or } B) = (\sim A) \text{ and } (\sim B)$
  (iv) $\sim(\forall x \in X \; : \; P(x)) = (\exists x \in X \; : \sim P(x))$
  (v) $\sim(\exists x \in X \; : \; P(x)) = (\forall x \in X \; : \sim P(x))$.

- Let A and B be statements. A *implies* B will be denoted by $A \Rightarrow B$. If A implies B, we take this to mean that if we wish to prove B, it suffices to prove A (A is a sufficient condition for B).

- The equivalence $A \Leftrightarrow B$ ("A and B are equivalent" or "A if and only if B," often written A iff B) of the statements A and B is defined by

$$(A \Leftrightarrow B) := (A \Rightarrow B) \text{ and } (B \Rightarrow A).$$

A is a necessary and sufficient condition for B, or vice versa.

- The statement $\sim B \Rightarrow \sim A$ is called the *contrapositive* of the statement $A \Rightarrow B$. In standard logic practices, any statement is considered equivalent to its contrapositive. It is often easier to prove a statement's contrapositive instead of directly proving the statement itself.

- To prove $A \Rightarrow B$ *by contradiction*, one supposes B is false (that $\sim B$ is true). Then, also assuming that A is true, one reaches a conclusion C which is already known to be false. This contradiction shows that if A is true $\sim B$ cannot be true, and hence B is true if A is true.

- Given two sets A and B, we define $A \cup B$ ("the union of A with B") as the set

$$A \cup B := \{x \: : \: x \in A \text{ or } x \in B \text{ or both}\}.$$

When speaking about unions, if we say $x \in A$ or $x \in B$ it also includes the possibility that x is in both A and B.

- We define $A \cap B$ ("the intersection of A with B") as the set

$$A \cap B := \{x \: : \: x \in A \text{ and } x \in B\}.$$

- Let A and B be subsets of X. Then
$$A \setminus B := \{x \in X : \ x \in A \text{ and } x \notin B\}$$
is the *relative complement* of B in A. When the set X is clear from the context we write also
$$A^c := X \setminus A$$
and call A^c the *complement* of A.

- If X is a set, then so is its *power set* $\mathcal{P}(X)$. The elements of $\mathcal{P}(X)$ are the subsets of X. Sometimes the power set is written 2^X for a reason which is made clear in Problem 2.8.

- Let $f : X \to Y$ be a function, then
$$im(f) := \{y \in Y; \ \exists x \in X : y = f(x)\}$$
is called the *image of f*. We say f is *surjective* (or onto) if $im(f) = Y$, *injective* (or one-to-one) if $f(x) = f(y)$ implies $x = y$ for all $x, y \in X$, and f is *bijective* if f is both injective and surjective.

- If X and Y are sets, the *Cartesian product* $X \times Y$ of X and Y is the set of all ordered pairs (x, y) with $x \in X$ and $y \in Y$.

- Let X be a set and $\mathcal{A} = \{A_i : \ i \in I\}$ be a family of sets and I is an index set. *Intersection and union of this family* are given by
$$\bigcap_{i \in I} A_i = \{x \in X; \forall i \in I : \ x \in A_i\}$$
and
$$\bigcup_{i \in I} A_i = \{x \in X; \exists i \in I : \ x \in A_i\}.$$

- Let $f : X \to Y$ be a function, and $A \subset X$ and $B \subset Y$ are subsets. *Image of A under f, $f(A)$* defined as
$$f(A) = \{f(x) \in Y : \ x \in A\}.$$

- *Inverse image of B under f* (or pre-image of B), $f^{-1}(B)$ defined as
$$f^{-1}(B) = \{x \in X : \ f(x) \in B\}.$$
Note that we can form $f^{-1}(B)$ for a set $B \subset Y$ even though f might not be one-to-one or onto.

- We will use standard notation, $\mathbb{N}$ for the set natural numbers, $\mathbb{Z}$ for the set of integers, $\mathbb{Q}$ for the set rational numbers, and $\mathbb{R}$ for the set real numbers. We have the natural containments:
$$\mathbb{N} \subset \mathbb{Z} \subset \mathbb{Q} \subset \mathbb{R}.$$

- Two sets A and B have the same *cardinality* if there is a bijection from A to B. In this case we write $A \sim B$. We say A is *countable* if $\mathbb{N} \sim A$. An infinite set that is not countable is called an uncountable set.

- *Schröder–Bernstein Theorem*: Assume that there exists one-to-one function $f : A \to B$ and another one-to-one function $g : B \to A$. Then there exists a one-to-one, onto function $h : A \to B$ and hence $A \sim B$.

Problem 1.1 Consider the four statements

$$(a) \quad \exists x \in \mathbb{R} \ \forall y \in \mathbb{R} \quad x + y > 0;$$
$$(b) \quad \forall x \in \mathbb{R} \ \exists y \in \mathbb{R} \quad x + y > 0;$$
$$(c) \quad \forall x \in \mathbb{R} \ \forall y \in \mathbb{R} \quad x + y > 0;$$
$$(d) \quad \exists x \in \mathbb{R} \ \forall y \in \mathbb{R} \quad y^2 > x.$$

1. Are the statements a, b, c, d true or false?

2. Find their negations.

Problem 1.2 Let $f : \mathbb{R} \to \mathbb{R}$. Find the negations of the following statements:

1. For any $x \in \mathbb{R}$ $f(x) \leq 1$.

2. The function f is increasing.

3. The function f is increasing and positive.

4. There exists $x \in \mathbb{R}^+$ such that $f(x) \leq 0$.

5. There exists $x \in \mathbb{R}$ such that for any $y \in \mathbb{R}$, if $x < y$ then $f(x) > f(y)$.

Problem 1.3 Replace ... by the appropriate quantifier: $\Leftrightarrow, \Leftarrow$, or $\Rightarrow$.

1. $x \in \mathbb{R}$ $x^2 = 4$ ...... $x = 2$;

2. $z \in \mathbb{C}$ $z = \overline{z}$ ...... $z \in \mathbb{R}$;

3. $x \in \mathbb{R}$ $x = \pi$ ...... $e^{2ix} = 1$.

Problem 1.4 Find the negation of: "Anyone living in Los Angeles who has blue eyes will win the Lottery and will take their retirement before the age of 50."

CHAPTER 1. ELEMENTARY LOGIC AND SET THEORY

Problem 1.5 Find the negation of the following statements:

1. Any rectangular triangle has a right angle.

2. In all the stables, the horses are black.

3. For any integer $x \in \mathbb{Z}$, there exists an integer $y \in \mathbb{Z}$ such that, for any $z \in \mathbb{Z}$, the inequality $z < x$ implies $z < x + 1$.

4. $\forall \varepsilon > 0 \; \exists \alpha > 0 \; / \; |x - 7/5| < \alpha \Rightarrow |5x - 7| < \varepsilon$.

Problem 1.6 Show that $\forall \varepsilon > 0 \; \exists N \in \mathbb{N}$ such that
$$(n \geq N \Rightarrow 2 - \varepsilon < \frac{2n+1}{n+2} < 2 + \varepsilon).$$

Problem 1.7 Let f, g be two functions defined from $\mathbb{R}$ into $\mathbb{R}$. Translate using quantifiers the following statements:

1. f is bounded above;
2. f is bounded;
3. f is even;
4. f is odd;
5. f is never equal to 0;
6. f is periodic;
7. f is increasing;
8. f is strictly increasing;
9. f is not the 0 function;
10. f does not have the same value at two different points;
11. f is less than g;
12. f is not less than g.

Problem 1.8 For two sets A and B show that the following statements are equivalent:

a) $A \subseteq B$
b) $A \cup B = B$
c) $A \cap B = A$

Problem 1.9 Establish the following set theoretic relations:

a) $A \cup B = B \cup A$, $\ A \cap B = B \cap A$ (Commutativity)

b) $A \cup (B \cup C) = (A \cup B) \cup C$, $\ A \cap (B \cap C) = (A \cap B) \cap C$ (Associativity)

c) $A \cup (B \cap C) = (A \cup B) \cap (A \cup C)$ and $A \cap (B \cup C) = (A \cap B) \cup (A \cap C)$ (Distributivity)

d) $A \subseteq B \ \Leftrightarrow \ B^c \subseteq A^c$

e) $A \backslash B = A \cap B^c$

f) $(A \cup B)^c = A^c \cap B^c$ and $(A \cap B)^c = A^c \cup B^c$ (De Morgan's laws)

Problem 1.10 Suppose the collection $\mathcal{B}$ is given by $\mathcal{B} = \{[1, 1 + \frac{1}{n}] : n \in \mathbb{N}\}$. Find $\bigcup_{B \in \mathcal{B}} B$ and $\bigcap_{B \in \mathcal{B}} B$.

Problem 1.11 Let A be a set and let $\mathcal{P}(A)$ denote the set of all subsets of A (i.e., the power set of A). Prove that A and $\mathcal{P}(A)$ do not have the same cardinality. (The term *cardinality* is used in mathematics to refer to the size of a set.)

Problem 1.12 If A and B are sets, then show that

a) $\mathcal{P}(A) \cup \mathcal{P}(B) \subseteq \mathcal{P}(A \cup B)$

b) $\mathcal{P}(A) \cap \mathcal{P}(B) = \mathcal{P}(A \cap B)$

Problem 1.13 Prove that for each nonempty set A, the function

$$\begin{aligned} f : \mathcal{P}(A) &\longrightarrow \{\chi_B\}_{B \in \mathcal{P}(A)}, \\ B &\longmapsto \chi_B \end{aligned}$$

is bijective. Here the characteristic function χ_B of B is defined as

$$\begin{aligned} \chi_B : A &\longrightarrow \{0, 1\}, \\ x &\longmapsto \begin{cases} 1 & \text{if } x \in B, \\ 0 & \text{if } x \in B^c. \end{cases} \end{aligned}$$

Problem 1.14 Give a necessary and sufficient condition for

$$A \times B = B \times A.$$

Problem 1.15 If A, B, C are sets, show that

a) $A \times B = \emptyset \Leftrightarrow A = \emptyset$ or $B = \emptyset$.

b) $(A \cup B) \times C = (A \times C) \cup (B \times C)$.

c) $(A \cap B) \times C = (A \times C) \cap (B \times C)$.

Problem 1.16 For an arbitrary function $f : X \longrightarrow Y$, prove that the following relations hold:

a) $f(\bigcup_{i \in I} A_i) = \bigcup_{i \in I} f(A_i)$.

b) $f(\bigcap_{i \in I} A_i) \subseteq \bigcap_{i \in I} f(A_i)$.

c) Give a counterexample to show that $f(\bigcap_{i \in I} A_i) = \bigcap_{i \in I} f(A_i)$ is not always true.

Problem 1.17 Suppose $f : A \to B$ and $g : B \to C$ are functions, show that

a) If both f and g are one-to-one, then $g \circ f$ is one-to-one.

b) If both f and g are onto, then $g \circ f$ is onto.

c) If both f and g are bijection, then $g \circ f$ is bijection.

Problem 1.18 For a function $f : X \longrightarrow Y$, show that the following statements are equivalent:

a) f is one-to-one.

b) $f(A \cap B) = f(A) \cap f(B)$ holds for all $A, B \in \mathcal{P}(X)$.

Problem 1.19 For an arbitrary function $f : X \longrightarrow Y$, prove the following identities:

a) $f^{-1}\left(\bigcup_{i \in I} B_i\right) = \bigcup_{i \in I} f^{-1}(B_i)$.

b) $f^{-1}\left(\bigcap_{i \in I} B_i\right) = \bigcap_{i \in I} f^{-1}(B_i)$.

c) $f^{-1}(B^c) = [f^{-1}(B)]^c$.

Problem 1.20 Show that

1. $\mathbb{N} \sim \mathbb{E}$.

2. $\mathbb{N} \sim \mathbb{Z}$.

3. $(-1, 1) \sim \mathbb{R}$.

Problem 1.21 Show that any nonempty subset of a countable set is finite or countable.

Problem 1.22 Let A be an infinite set. Show that A is countable if and only if there exists $f : A \to \mathbb{N}$ which is 1-to-1. Use this to prove that $\mathbb{Z}$, $\mathbb{N} \times \mathbb{N}$, $\mathbb{N}^r$, for any $r \geq 1$, and $\mathbb{Q}$ are countable.

Problem 1.23 Show that the countable union of finite or countable sets is countable.

Problem 1.24 An algebraic number is a root of a polynomial, whose coefficients are rational. Show that the set of all algebraic numbers is countable.

Problem 1.25 Show that the set $\mathbb{R}$ is uncountable.

Problem 1.26 The power set of $\mathbb{N}$, i.e., $\mathcal{P}(\mathbb{N})$, is not countable as well as the sets $\mathbb{R}$, and $\{0,1\}^{\mathbb{N}}$ the set of all the sequences which takes values 0 or 1. Use this to show that the set of all irrationals is not countable.

Problem 1.27 Let A and B be two nonempty sets. Assume there exist $f : A \to B$ and $g : B \to A$ which are 1-to-1 (or injective). Then there exists a bijection $h : A \to B$.
This conclusion is known as the Schröder–Bernstein theorem.

Solutions

Solution 1.1

1. (a) is false. Since its negation $\forall x \in \mathbb{R} \; \exists y \in \mathbb{R} \quad x + y \leq 0$ is true. Because if $x \in \mathbb{R}$, there exists $y \in \mathbb{R}$ such that $x + y \leq 0$. For example, we may take $y = -(x+1)$ which gives $x + y = x - x - 1 = -1 \leq 0$.

2. (b) is true. Indeed for $x \in \mathbb{R}$, one can take $y = -x + 1$ which gives $x + y = 1 > 0$. The negation of (b) is $\exists x \in \mathbb{R} \; \forall y \in \mathbb{R} \quad x + y \leq 0$.

3. (c) : $\forall x \in \mathbb{R} \; \forall y \in \mathbb{R} \quad x + y > 0$ is false. Indeed one may take $x = -1$, $y = 0$. The negation of (c) is $\exists x \in \mathbb{R} \; \exists y \in \mathbb{R} \; x + y \leq 0$.

4. (d) is true. Indeed one may take $x = -1$. The negation is: $\forall x \in \mathbb{R} \; \exists y \in \mathbb{R} \quad y^2 \leq x$.

Solution 1.2

1. This statement may be rewritten as: (For every $x \in \mathbb{R}$) $(f(x) \leq 1)$. The negation of "(For every $x \in \mathbb{R}$)" is "There exists $x \in \mathbb{R}$" and the negation of "$(f(x) \leq 1)$" is "$f(x) > 1$." Hence the negation of the statement is: "There exists $x \in \mathbb{R}, f(x) > 1$."

2. First let us rewrite the statement "The function f is increasing": "for any real numbers (x_1, x_2), if $x_1 \leq x_2$ then $f(x_1) \leq f(x_2)$." This may be rewritten as: "(for any real numbers x_1 and x_2) $(x_1 \leq x_2$ implies $f(x_1) \leq f(x_2))$." The negation of the first part is: "(there exists a pair of real numbers (x_1, x_2))" and the negation of the second part is: "$(x_1 \leq x_2$ and $f(x_1) > f(x_2))$". Hence the negation of the complete statement is: "There exist $x_1 \in \mathbb{R}$ and $x_2 \in \mathbb{R}$ such that $x_1 \leq x_2$ and $f(x_1) > f(x_2)$."

3. The negation is: the function f is not increasing or is not positive. We already did describe the statement "the function f is not increasing." Let us focus on "the function f is not positive." We get: "there exists $x \in \mathbb{R}, f(x) < 0$." Therefore the negation of the complete statement is: "there exist $x_1 \in \mathbb{R}$ and $x_2 \in \mathbb{R}$ such that $x_1 < x_2$ and $f(x_1) \geq f(x_2)$, or there exists $x \in \mathbb{R}, f(x) < 0$."

4. This statement may be rewritten as follows: "(there exists $x \in \mathbb{R}^+$) $(f(x) \leq 0)$." The negation of the first part is: "(for any $x \in \mathbb{R}^+$)," and for the second part: "$(f(x) > 0)$." Hence the negation of the complete statement is: "for any $x \in \mathbb{R}^+$, $f(x) > 0$."

5. This statement may be rewritten as follows: "$(\exists x \in \mathbb{R})(\forall y \in \mathbb{R})(x < y \Rightarrow f(x) > f(y))$." The negation of the first part is: "$(\forall x \in \mathbb{R})$," for the second part: "$(\exists y \in \mathbb{R})$," and for the third part: "$(x < y$ and $f(x) \leq f(y))$." Hence the negation of the complete statement is: "$\forall x \in \mathbb{R}, \exists y \in \mathbb{R}, x < y$ and $f(x) \leq f(y)$."

Solution 1.3

1. $\Leftarrow$
2. $\Leftrightarrow$
3. $\Rightarrow$

Solution 1.4

"There exists one person living in Los Angeles who has blue eyes who will not win the Lottery or will retire after the age of 50."

Solution 1.5

1. A triangle with no right angle, is not rectangular.

2. There exists a stable in which there exists at least one horse who is not black.

3. If we rewrite the statement in mathematical language:
$$\forall x \in \mathbb{Z} \ \exists y \in \mathbb{Z} \ \forall z \in \mathbb{Z} \quad (z < x \quad \Leftrightarrow \quad z < x+1),$$
the negation is
$$\exists x \in \mathbb{Z} \ \forall y \in \mathbb{Z} \ \exists z \in \mathbb{Z} \quad (z < x \text{ and } z \geq x+1).$$

4. $\exists \varepsilon > 0 \ \forall \alpha > 0 \quad (|x - 7/5| < \alpha \text{ and } |5x - 7| \geq \varepsilon).$

Solution 1.6

First note that for $n \in \mathbb{N}$, $\dfrac{2n+1}{n+2} \leq 2$ since $2n+1 \leq 2(n+2)$. Let $\varepsilon > 0$, we have
$$\forall n \in \mathbb{N} \quad \frac{2n+1}{n+2} < 2 + \varepsilon;$$
let us find a condition on n such that the inequality
$$2 - \varepsilon < \frac{2n+1}{n+2}$$
is true. We have
$$2 - \varepsilon < \frac{2n+1}{n+2} \Leftrightarrow (2-\varepsilon)(n+2) < 2n+1$$
$$\Leftrightarrow 3 < \varepsilon(n+2)$$
$$\Leftrightarrow n > \frac{3}{\varepsilon} - 2.$$

Here ε is given, let us pick $N \in \mathbb{N}$ such that $N > \dfrac{3}{\varepsilon} - 2$. Hence, for any $n \geq N$ we have $n \geq N > \dfrac{3}{\varepsilon} - 2$. Consequently $2 - \varepsilon < \dfrac{2n+1}{n+2}$. As a conclusion, for any $\varepsilon > 0$, we found $N \in \mathbb{N}$ such that for any $n \geq N$ we have $2 - \varepsilon < \dfrac{2n+1}{n+2}$ and $\dfrac{2n+1}{n+2} < 2 + \varepsilon$.

Solution 1.7

1. $\exists M \in \mathbb{R} \ \forall x \in \mathbb{R} \ f(x) \leq M$;

2. $\exists M \in \mathbb{R} \ \exists m \in \mathbb{R} \ \forall x \in \mathbb{R} \ m \leq f(x) \leq M$;

3. $\forall x \in \mathbb{R} \ f(x) = f(-x)$;

4. $\forall x \in \mathbb{R} \ f(x) = -f(-x)$;

5. $\forall x \in \mathbb{R} \ f(x) \neq 0$;

6. $\exists a \in \mathbb{R}^* \ \forall x \in \mathbb{R} \ f(x+a) = f(x)$;

7. $\forall (x,y) \in \mathbb{R}^2 \ (x \leq y \Rightarrow f(x) \leq f(y))$;

8. $\forall (x,y) \in \mathbb{R}^2 \ (x \leq y \Rightarrow f(x) > f(y))$;

9. $\exists x \in \mathbb{R} \ f(x) \neq 0$;

10. $\forall (x,y) \in \mathbb{R}^2 \ (x \neq y \Rightarrow f(x) \neq f(y))$;

11. $\forall x \in \mathbb{R} \ f(x) \leq g(x)$;

12. $\exists x \in \mathbb{R} \ f(x) > g(x)$.

Solution 1.8

$(a \Rightarrow b)$

Suppose $A \subseteq B$. Let $x \in A \cup B$, then $x \in A$ or $x \in B$. If $x \in A$, then since $A \subseteq B$, we have $x \in B$. Thus, for any $x \in A \cup B$, $x \in B$, so $A \cup B \subseteq B$. Let $x \in B$, then $x \in A \cup B$ so $B \subseteq A \cup B$, and hence $A \cup B = B$.

$(b \Rightarrow c)$

Let $x \in A \cap B \Rightarrow x \in A$ so $A \cap B \subseteq A$. Let $x \in A \Rightarrow x \in A \cup B = B$ so $x \in B$ and therefore $x \in A \cap B$ so $A \subseteq A \cap B$. Thus $A \cap B = A$.

$(c \Rightarrow a)$

Let $x \in A$, then by hypothesis $x \in A \cap B$, which in turn implies that $x \in B$ as well. Thus $A \subseteq B$.

Solution 1.9

a) Follows directly from definitions.

b) Follows directly from definitions.

c) To establish this equality, note the following:

$$\begin{aligned}
x \in A \cup (B \cap C) &\Leftrightarrow x \in A \text{ or } x \in B \cap C \\
&\Leftrightarrow x \in A \text{ or } (x \in B \text{ and } x \in C) \\
&\Leftrightarrow (x \in A \text{ or } x \in B) \text{ and } (x \in A \text{ or } x \in C) \\
&\Leftrightarrow x \in A \cup B \text{ and } x \in A \cup C \\
&\Leftrightarrow x \in (A \cup B) \cap (A \cup C).
\end{aligned}$$

Similarly,

$$\begin{aligned}
x \in A \cap (B \cup C) &\Leftrightarrow x \in A \text{ and } x \in B \cup C \\
&\Leftrightarrow x \in A \text{ and } (x \in B \text{ or } x \in C) \\
&\Leftrightarrow (x \in A \text{ and } x \in B) \text{ or } (x \in A \text{ and } x \in C) \\
&\Leftrightarrow x \in A \cap B \text{ or } x \in A \cap C \\
&\Leftrightarrow x \in (A \cap B) \cup (A \cap C).
\end{aligned}$$

d) Let $A \subseteq B$, then $x \in B^c \Rightarrow x \notin B$ and so $x \notin A$ (i.e., $x \in A^c$), therefore, $B^c \subseteq A^c$. Conversely, if $B^c \subseteq A^c$ is true, then by the preceding case, $A = (A^c)^c \subseteq (B^c)^c = B$.

e) Note that

$$\begin{aligned}
x \in A \backslash B &\Leftrightarrow x \in A \text{ and } x \notin B \\
&\Leftrightarrow x \in A \text{ and } x \in B^c \\
&\Leftrightarrow x \in A \cap B^c.
\end{aligned}$$

f) Note that

$$\begin{aligned}
x \in (A \cap B)^c &\Leftrightarrow x \notin A \cap B \\
&\Leftrightarrow x \notin A \text{ or } x \notin B \\
&\Leftrightarrow x \in A^c \text{ or } x \in B^c \\
&\Leftrightarrow x \in A^c \cup B^c.
\end{aligned}$$

Similarly,

$$\begin{aligned}
x \in (A \cup B)^c &\Leftrightarrow x \notin A \cup B \\
&\Leftrightarrow x \notin A \text{ and } x \notin B \\
&\Leftrightarrow x \in A^c \text{ and } x \in B^c \\
&\Leftrightarrow x \in A^c \cap B^c.
\end{aligned}$$

Solution 1.10

Clearly, $\bigcup_{B \in \mathcal{B}} B = [1,2]$ and $\bigcap_{B \in \mathcal{B}} B = \{1\}$.

Solution 1.11

This is a proof by contradiction. Suppose they have the same cardinality, then there exists a bijection
$$T : A \to \mathcal{P}(A).$$
Let $K = \{x \in A : x \notin T(x)\}$. Since T is onto, there exists $y \in A$ such that $T(y) = K$. If $y \in K$, then by the definition of K we can conclude $y \notin T(y) = K$, so $y \notin K$. Similarly, if $y \notin K$, $y \notin T(y)$ so we conclude that $y \in K$. In both cases we have reached a contradiction.

Solution 1.12

a)
$$\begin{aligned} X \in \mathcal{P}(A) \cup \mathcal{P}(B) &\Rightarrow X \subseteq A \text{ or } X \subseteq B \\ &\Rightarrow X \subseteq A \cup B \\ &\Rightarrow X \in \mathcal{P}(A \cup B). \end{aligned}$$

b)
$$\begin{aligned} X \in \mathcal{P}(A) \cap \mathcal{P}(B) &\Leftrightarrow X \subseteq A \text{ and } X \subseteq B \\ &\Leftrightarrow X \subseteq A \cap B \\ &\Leftrightarrow X \in \mathcal{P}(A \cap B). \end{aligned}$$

Solution 1.13

Clearly, for any $B \in \mathcal{P}(A)$, $f(B) = \chi_B$. Note that f is one-to-one and onto.

Solution 1.14

Clearly, $A \times B = B \times A \Leftrightarrow A = B$.

Solution 1.15

a) Without loss of generality, let $A \neq \emptyset$ then if
$$\begin{aligned} B \neq \emptyset &\Leftrightarrow (\exists x \in A \text{ and } \exists y \in B) \\ &\Leftrightarrow (x,y) \in A \times B \\ &\Leftrightarrow A \times B \neq \emptyset. \end{aligned}$$

b) Suppose $(x,y) \in (A \cup B) \times C$ then $y \in C$ and $x \in A$ or $x \in B$. So $(x,y) \in A \times C$ or $(x,y) \in B \times C$, thus $(x,y) \in (A \times C) \cup (B \times C)$ and $(A \cup B) \times C \subseteq (A \times C) \cup (B \times C)$. Conversely, if $(x,y) \in (A \times C) \cup (B \times C)$, then $(x,y) \in A \times C$ or $(x,y) \in B \times C$, which means that $y \in C$ and $x \in A$ or $x \in B$; therefore, $(x,y) \in (A \cup B) \times C$ and $(A \times C) \cup (B \times C) \subseteq (A \cup B) \times C$.

c) Suppose $(x,y) \in (A \cap B) \times C$ then $y \in C$ and $x \in A$ and $x \in B$. So $(x,y) \in A \times C$ and $(x,y) \in B \times C$, thus $(x,y) \in (A \times C) \cap (B \times C)$ and $(A \cap B) \times C \subseteq (A \times C) \cap (B \times C)$. Conversely, if $(x,y) \in (A \times C) \cap (B \times C)$, then $(x,y) \in A \times C$ and $(x,y) \in B \times C$, which means that $y \in C$ and $x \in A$ and $x \in B$; therefore, $(x,y) \in (A \cap B) \times C$ and $(A \times C) \cap (B \times C) \subseteq (A \cap B) \times C$.

Solution 1.16

a) Note the following:

$$
\begin{aligned}
y \in f\left(\bigcup_{i \in I} A_i\right) &\Leftrightarrow \exists x \in \bigcup_{i \in I} A_i \text{ with } y = f(x) \\
&\Leftrightarrow \exists i \in I \text{ with } x \in A_i \text{ and } y = f(x) \\
&\Leftrightarrow \exists i \in I \text{ with } y \in f(A_i) \\
&\Leftrightarrow y \in \bigcup_{i \in I} f(A_i).
\end{aligned}
$$

b) Since $\bigcap_{i \in I} A_i \subseteq A_i$ we have $f(\bigcap_{i \in I} A_i) \subseteq f(A_i)$ for each i. We obtain that

$$f\left(\bigcap_{i \in I} A_i\right) \subseteq \bigcap_{i \in I} f(A_i).$$

c) Let $A_1 = \{0\}$ and $A_2 = \{1\}$ and $X = Y = \{0,1\}$. Define $f : X \longrightarrow Y$ by $f(0) = f(1) = 0$, then $f(A_1) = \{0\} = f(A_2)$. Therefore, $f(A_1) \cap f(A_2) = \{0\}$, while $A_1 \cap A_2 = \emptyset$ and $f(A_1 \cap A_2) = \emptyset$.

Solution 1.17

a) Let $f : A \to B$ and $g : B \to C$ are given functions. Let $x_1, x_2 \in A$ with $x_1 \neq x_2$, since f is 1-1, we know that $f(x_1) \neq f(x_2)$. So now we have two distinct points $f(x_1)$ and $f(x_2)$ in B. Since g is 1-1, we also have $g(f(x_1)) \neq g(f(x_2))$. This is same as $(g \circ f)(x_1) \neq (g \circ f)(x_2)$.

b) Suppose $c \in C$, since g is onto, we know that there is a $b \in B$ with $g(b) = c$. Furthermore since f is also onto, there is some $a \in A$ with $f(a) = b$. But this means

$$(g \circ f)(a) = g(f(a)) = g(b) = c$$

so we have the required $a \in A$.

c) Since both f and g are bijective, they are both 1-1 and onto. So $g \circ f$ is 1-1 by part a) above and $g \circ f$ is onto by part b) above. Therefore $g \circ f$ is 1-1 and onto, in other words $g \circ f$ is a bijection.

Solution 1.18

$(a \Rightarrow b)$ If $y \in f(A) \cap f(B)$, then $\exists a \in A$ and $b \in B$ such that $y = f(a) = f(b)$. Since f is one-to-one we know $a = b \in A \cap B$ and therefore, $y \in f(A \cap B)$. Thus $f(A) \cap f(B) \subseteq f(A \cap B)$, but from the problem above $f(A \cap B) \subseteq f(A) \cap f(B)$.

$(b \Rightarrow a)$ If b holds, notice that if A and B were disjoint subsets of X, we then have $f(A) \cap f(B) = \emptyset$. Now let $f(a) = f(b)$. Then let $A = \{a\}$ and $B = \{b\}$. Thus $f(A \cap B) = f(A) \cap f(B) = f(a) \cap f(b) = f(a) \cap f(a) = f(a) \neq \emptyset$. So $A \cap B \neq \emptyset$, and therefore, $a = b$.

Solution 1.19

a)
$$\begin{aligned}
x \in f^{-1}\left(\bigcup_{i \in I} B_i\right) &\Leftrightarrow f(x) \in \bigcup_{i \in I} B_i \\
&\Leftrightarrow \exists i \in I \text{ such that } f(x) \in B_i \\
&\Leftrightarrow \exists i \in I \text{ such that } x \in f^{-1}(B_i) \\
&\Leftrightarrow x \in \bigcup_{i \in I} f^{-1}(B_i).
\end{aligned}$$

b)
$$\begin{aligned}
x \in f^{-1}\left(\bigcap_{i \in I} B_i\right) &\Leftrightarrow f(x) \in \bigcap_{i \in I} B_i \\
&\Leftrightarrow f(x) \in B_i \quad \forall i \in I \\
&\Leftrightarrow x \in f^{-1}(B_i) \quad \forall i \in I \\
&\Leftrightarrow x \in \bigcap_{i \in I} f^{-1}(B_i).
\end{aligned}$$

c)
$$\begin{aligned}
x \in f^{-1}(B^c) &\Leftrightarrow f(x) \in B^c \\
&\Leftrightarrow f(x) \notin B \\
&\Leftrightarrow x \notin f^{-1}(B) \\
&\Leftrightarrow x \in \left[f^{-1}(B)\right]^c.
\end{aligned}$$

Solution 1.20

1. Let $f : \mathbb{N} \to \mathbb{E}$ given by $f(n) = 2n$. Clearly this map is one-to-one and onto.

2. This can be shown by defining a bijection $f : \mathbb{N} \to \mathbb{Z}$ as

$$f(n) = \begin{cases} \frac{1-n}{2} & \text{if } n \text{ is odd,} \\ \frac{n}{2} & \text{if } n \text{ is even.} \end{cases}$$

3. Using calculus one can show that the function

$$f : (-1, 1) \to \mathbb{R}$$

defined by

$$f(x) = \frac{x}{x^2 - 1}$$

is one-to-one and onto. In fact, $(a, b) \sim \mathbb{R}$ for any interval (a, b).

Solution 1.21

Let A be a countable set and $B \subset A$. Without loss of generality assume B not empty and not finite. Let us prove that B is countable. Since A is countable, there exists a bijection $f : A \to \mathbb{N}$. Consider the restriction of f to B, denoted by $f_B : B \to \mathbb{N}$. f_B is a bijection from B into $f(B) \subset \mathbb{N}$. Clearly $f(B)$ is not empty and is not finite. Let us prove that $f(B)$ is in bijection with $\mathbb{N}$. Indeed set $b_0 = \min f(B)$. Then define $b_1 = \min f(B) \setminus \{b_0\}$. Once b_n is built, we define $b_{n+1} = \min f(B) \setminus \{b_0, b_1, \ldots, b_n\}$. By induction, we build the set $\{b_n, n \in \mathbb{N}\} \subset f(B)$ where $b_n < b_{n+1}$, for $n \in \mathbb{N}$. Assume $f(B) \setminus \{b_n, n \in \mathbb{N}\} \neq \emptyset$. Let $b \in f(B) \setminus \{b_n, n \in \mathbb{N}\}$. Then we have $b_n < b$ for any $n \in \mathbb{N}$. This is a contradiction since any increasing sequence of elements in $\mathbb{N}$ is not bounded above. So $f(B) = \{b_n, n \in \mathbb{N}\}$. Define $g : f(B) \to \mathbb{N}$ by $g(b_n) = n$, for any $n \in \mathbb{N}$. g is a bijection. Clearly $g \circ f_B : B \to \mathbb{N}$ is a bijection.

Solution 1.22

It is clear that if A is countable, then there exists a bijection $f : A \to \mathbb{N}$ which is also 1-to-1. So assume there exists $f : A \to \mathbb{N}$ which is 1-to-1. Let us prove that A is countable. Since A is infinite and f is 1-to-1, then $f(A)$ is infinite. In the previous problem, we showed that $f(A)$ is countable. Since f restricted to A into $f(A)$ is a bijection, then one can construct a bijection from A into $\mathbb{N}$, i.e. A is countable. Let us complete the proof by showing that $\mathbb{Z}$, $\mathbb{N}^r$, for any $r \geq 1$, and $\mathbb{Q}$ are countable. Note that $\mathbb{N} \times \mathbb{N} = \mathbb{N}^r$ for $r = 2$. The map $f : \mathbb{Z} \to \mathbb{N}$ defined by

$$f(n) = \begin{cases} 2n + 1 & \text{if } n \in \mathbb{N}, \\ -2n & \text{if } n \notin \mathbb{N}, \end{cases}$$

is 1-to-1. The first part shows that $\mathbb{Z}$ is countable. In order to show that $\mathbb{N}^r$, for $r \geq 1$, is countable, consider the set of prime numbers $\mathcal{P}$. We know that $\mathcal{P}$ is infinite. So for any $r \geq 1$, consider any subset $\mathcal{P}_r$ of $\mathcal{P}$ with r elements. And write $\mathcal{P}_r = \{p_1, \ldots, p_r\}$. Define the map $f : \mathbb{N}^r \to \mathbb{N}$ by

$$f(n_1, \ldots, n_r) = p_1^{n_1} \times \cdots \times p_r^{n_r}.$$

The elementary number theory shows that f is 1-to-1. Hence the first part shows that $\mathbb{N}^r$ is countable. Finally define $f : \mathbb{Q} \to \mathbb{N}$ by

$$f\left(\frac{n}{m}\right) = 2^{\operatorname{sign}(n)} \times 3^{|n|} \times 5^m.$$

It is easy to check that f is 1-to-1. Hence $\mathbb{Q}$ is countable.

Solution 1.23

Let $\{A_i\}_{i \in I}$ be a family of subsets of a set X such that A_i is finite or countable for any $i \in I$ and I is countable. Without loss of generality assume $I = \mathbb{N}$. Set $A = \bigcup_{n \in \mathbb{N}} A_n$. Let us prove that A is countable. Without loss of generality, assume A is not finite. For any $a \in A$, set $n_a = \min\{n \in \mathbb{N}; a \in A_n\}$. Since A_n is finite or countable, for any $n \in \mathbb{N}$, there exists $f_n : A_n \to \mathbb{N}$ which is 1-to-1. Define $f : A \to \mathbb{N}$ by $f(a) = f_{n_a}(a)$. Then it is easy to check that f is 1-to-1. This proves that A is countable.

Solution 1.24

Let $P_n[x]$ be the set of polynomial functions with rational coefficients. Obviously there is a bijection from $P_n[x]$ into $\mathbb{Q}^n$. Since $\mathbb{Q}$ is countable, there exists a bijection $f : \mathbb{Q} \to \mathbb{N}$. Hence the map $F : \mathbb{Q}^n \to \mathbb{N}^n$ defined by $F(r_1, \ldots, r_n) = (f(r_1), \ldots, f(r_n))$ is a bijection. Since $\mathbb{N}^n$ is countable, we conclude that $\mathbb{Q}^n$ is countable and consequently $P_n[x]$ is countable. The set

$$R_n = \bigcup_{P \in P_n[x]} \{x \in \mathbb{R}; P(x) = 0\}$$

is countable since it is a countable union of finite sets. Note that R_n is infinite since $\mathbb{Q} \subset R_n$, for any $n \geq 1$. Since the set of algebraic numbers $\mathcal{A}(\mathbb{R})$ is given by

$$\mathcal{A}(\mathbb{R}) = \bigcup_{n \geq 1} R_n,$$

then it is countable being a countable union of countable sets.

Solution 1.25

First notice that $(0,1) \sim \mathbb{R}$, because the function defined by

$$f : (0,1) \to \mathbb{R}$$

defined by

$$f(x) = \tan\left(\pi x - \frac{1}{2}\right)$$

is one-to-one and onto. Next we claim that $(0,1)$ is uncountable. The proof of this claim uses a *diagonalization* argument due to Cantor. Suppose to the contrary that $(0,1)$ is countable. Then all real numbers in $(0,1)$ can be written as an *exhaustive* list $x_1, x_2, x_3, \ldots, x_k, \ldots$ where each x_k is given as a decimal expansion. There are certain real numbers in $(0,1)$ that have two decimal expansions. For example, $\frac{1}{10}$ has the two representations

$$0.10000 \cdots \text{ and } 0.09999 \cdots.$$

We can give preference to one of the representations, but it is not necessary to do so as can be seen in the following argument. Suppose all the real numbers in $(0,1)$ are given by the following list:

$$x_1 = 0.a_{11}a_{12}a_{13}a_{14}\cdots$$
$$x_2 = 0.a_{21}a_{22}a_{23}a_{24}\cdots$$
$$x_3 = 0.a_{31}a_{32}a_{33}a_{34}\cdots$$
$$x_4 = 0.a_{41}a_{42}a_{43}a_{44}\cdots$$
$$\cdots$$
$$x_k = 0.a_{k1}a_{k2}a_{k3}a_{k4}\cdots$$
$$\cdots$$

Our goal is to write down another real number y in $(0,1)$ which does not appear in the above list. Now let b_1 be a digit different from $0, a_{11}$, and 9; b_2 be a digit different from $0, a_{22}$, and 9; b_3 be a digit different from $0, a_{33}$, and 9; etc. Consider a number y with decimal representation

$$y = 0.b_1b_2b_3b_4\cdots$$

clearly $y \in (0,1)$, furthermore y is not one of the numbers with two decimal representations, since $b_n \neq 0, 9$. Moreover $y \neq x_k$ for any k because the kth digit in the decimal representation for y and x_k are different. Therefore there is no list of *all* real numbers in $(0,1)$, and thus $(0,1)$ is not countable. Since $(0,1) \sim \mathbb{R}$, $\mathbb{R}$ is uncountable too.

Solution 1.26

We have seen that for any set X, there does not exist an onto map from X into the power set $\mathcal{P}(X)$. Hence $\mathcal{P}(\mathbb{N})$ is an infinite set which is not in bijection with $\mathbb{N}$. Hence $\mathcal{P}(\mathbb{N})$ is not countable. First let us prove that $\{0,1\}^{\mathbb{N}}$ is not countable. Assume not, then there exists a bijection $f : \mathbb{N} \to \{0,1\}^{\mathbb{N}}$. Set $\varepsilon = (\varepsilon_n) \in \{0,1\}^{\mathbb{N}}$ defined by $\varepsilon_n = 1 - f(n)_n$, where $f(n)_n$ is the nth term in the sequence $f(n) \in \{0,1\}^{\mathbb{N}}$. Obviously $\varepsilon \neq f(n)$, for any $n \in \mathbb{N}$. Hence ε does not belong to the range of f contradicting the onto behavior of f. Hence $\{0,1\}^{\mathbb{N}}$ is not countable. In order to prove that $\mathbb{R}$ is not countable, we find a subset of $\mathbb{R}$ which is not countable. Indeed, consider the set

$$C = \left\{\sum_{n=0}^{\infty} \frac{\varepsilon_n}{3^n}; (\varepsilon_n) \in \{0,1\}^{\mathbb{N}}\right\}.$$

C is the Cantor triadic set. It is clear that C is a subset of $[0,1] \subset \mathbb{R}$, and it is in bijection with $\{0,1\}^{\mathbb{N}}$. Hence C is not countable. Hence $\mathbb{R}$ is not countable. Since $\mathbb{R}$ is not countable and $\mathbb{Q}$ is countable, $\mathbb{R} \setminus \mathbb{Q}$ is not countable since the union of two countable sets is countable. Clearly the set $\mathbb{R} \setminus \mathbb{Q}$ is the set of all irrationals.

Solution 1.27

Let us first prove that if there exists a 1-to-1 map $f : A \to B$ with $B \subset A$, then there exists a bijection $h : A \to B$. Indeed set $B_0 = A \setminus B$, and $B_{n+1} = f(B_n)$. Note that the family $\{B_n\}$ is pairwise disjoint, i.e., $B_n \cap B_m = \emptyset$, for any $n \neq m$. Indeed we have $B_0 \cap B = \emptyset$ and $B_n \subset B$ for all $n \geq 1$. Hence $B_0 \cap B_n = \emptyset$ for any $n \geq 1$. Since f is 1-to-1, we get $f^m(B_0) \cap f^m(B_n) = \emptyset$, for any $n, m \in \mathbb{N}$. In other words we have $B_n \cap B_{n+m} = \emptyset$, for any $n, m \in \mathbb{N}$. This proves our claim. It is clear that we have $A \setminus \left(\bigcup_{n \geq 0} B_n \right) \subset B$. Indeed, if $a \in A \setminus \left(\bigcup_{n \geq 0} B_n \right)$, then $a \in A \setminus B_0 = B$. Define $h : A \to B$ by

$$h(a) = \begin{cases} f(a) & \text{if } a \in \bigcup_{n \geq 0} B_n, \\ a & \text{if } a \notin \bigcup_{n \geq 0} B_n. \end{cases}$$

We claim that h is a bijection. Indeed, it is straightforward that h is 1-to-1 since f is 1-to-1. Let us prove that h is onto (or surjective). Let $y \in B$. If $y \notin \bigcup_{n \geq 0} B_n$, then we have $h(y) = y$, i.e., y is in the range of h. Assume $y \in \bigcup_{n \geq 0} B_n$. Then there exists $n \geq 1$ such that $y \in B_n$. Note that $y \notin B_0$ because $y \in B$. Since $B_n = f(B_{n-1})$, then there exists $a \in B_{n-1}$ such that $f(a) = y$. But $f(a) = h(a)$. Hence y is in the range of h. This completes the proof that h is a bijection. In the general case, we do not assume $B \subset A$, but we do assume the existence of $f : A \to B$ and $g : B \to A$ which are 1-to-1. Clearly $g \circ f : A \to g(B)$ is 1-to-1, and $g(B) \subset A$. The first part of our proof shows the existence of a bijection $h_A : A \to g(B)$. Note that the restriction of g from B into $g(B)$ is a bijection. The map $g^{-1} \circ h_A : A \to B$ is a bijection.

Chapter 2

Real Numbers

Ah, but my Computations, People say,
Have Squared the Year to human Compass, eh?
If so, by striking from the Calendar
Unborn Tomorrow, and dead Yesterday

Omar Khayyam (1048–1123)

- *Mathematical induction* is a method of proof used to establish that a given statement is true for all natural numbers. Let $S(n)$ be a statement about the positive integer n. If

  1. $S(1)$ is true and
  2. for all $k \geq 1$, the truth of $S(k)$ implies the truth of $S(k+1)$,

  then $S(n)$ is true for all $n \geq 1$.
  Verifying $S(1)$ is true is called the *basis step*. The assumption that $S(k)$ is true for some $k \geq 1$ is called the *induction hypothesis*. Using the induction hypothesis to prove $S(k+1)$ is true is called the *induction step*. There are variants of mathematical induction used in practice, for example if one wants to prove a statement not for all natural numbers but only for all numbers greater than or equal to a certain number b, then

  1. Show $S(b)$ is true.
  2. Assume $S(m)$ is true for $m \geq b$ and show that truth of $S(m)$ implies the truth of $S(m+1)$.

  Another generalization, called *strong induction*, says that in the *inductive step* we may assume not only the statement holds for $S(k+1)$ but also that it is true for $S(m)$ for all $m \leq k+1$. In strong induction it is not necessary to list the *basis step*, it is clearly true that the statement holds for all previous cases. The *inductive step* of a strong induction in this case corresponds to the *basis step* in ordinary induction.

- Let A be a nonempty subset of $\mathbb{R}$. The number b is called an *upper bound* for A if for all $x \in A$, we have $x \leq b$. A number b is called a *least upper bound* of A if, first, b is an upper bound for A and, second, b is less than or equal to every upper bound for A. The *supremum* of A (also called least upper bound of A) is denoted by $\sup(A)$, $\sup A$ or $lub(A)$. If $A \subset \mathbb{R}$ is not bounded above, we say that $\sup A$ is infinite and write $\sup A = +\infty$.

- A *lower bound* for a set $A \subset \mathbb{R}$ is a number b such that $b \leq x$ for all $x \in A$. Also b is called a *greatest lower bound* if and only if it is a lower bound and for any lower bound c of A, $c \leq b$. The *infimum* of A (also called greatest lower bound of A) is denoted by $\inf(A)$, $\inf A$ or $glb(A)$. If $A \subset \mathbb{R}$ is not bounded below, we set $\inf A = -\infty$.

- *Well-Ordering Property*: If A is a nonempty subset of $\mathbb{N}$, then there is a smallest element in A, i.e., there is an $a \in A$ such that $a \leq x$ for every $x \in A$.

- *Archimedean Property*: If $x \in \mathbb{Q}$, then there is an integer n with $x < n$.

- For each $n \in \mathbb{N}$, let I_n be a nonempty closed interval in $\mathbb{R}$. The family $\{I_n : n \in \mathbb{N}\}$ is called a *nest of intervals* if the following conditions hold:
  - $I_{n+1} \subset I_n$ for all $n \in \mathbb{N}$.
  - For each $\varepsilon > 0$, there is some $n \in \mathbb{N}$ such that $|I_n| < \varepsilon$.

Problem 2.1 Prove that $\sqrt{2}$ is not rational.

Problem 2.2 Show that two real numbers x and y are equal if and only if $\forall \varepsilon > 0$ it follows that $|x - y| < \varepsilon$.

Problem 2.3 Use the induction argument to prove that
$$1 + 2 + \cdots + n = \frac{n(n+1)}{2}$$
for all natural numbers $n \geq 1$.

Problem 2.4 Use the induction argument to prove that
$$1^2 + 2^2 + \cdots + n^2 = \frac{2n^3 + 3n^2 + n}{6}$$
for all natural numbers $n \geq 1$.

CHAPTER 2. REAL NUMBERS

Problem 2.5 Use the induction argument to prove that $n^3 + 5n$ is divisible by 6 for all natural numbers $n \geq 1$.

Problem 2.6 Use induction to prove that if $1 + x > 0$, then $(1+x)^n \geq 1 + nx$ for all natural numbers $n \geq 0$. This is known as Bernoulli's inequality.

Problem 2.7 Consider the Fibonacci numbers $\{F_n\}$ defined by
$$F_1 = 1, F_2 = 1, \text{ and } F_{n+2} = F_{n+1} + F_n.$$
Show that
$$F_n = \frac{(1+\sqrt{5})^n - (1-\sqrt{5})^n}{2^n \sqrt{5}}, n = 1, 2, \ldots.$$

Problem 2.8 Show by induction that if X is a finite set with n elements, then $\mathcal{P}(X)$, the power set of X (i.e., the set of subsets of X), has 2^n elements.

Problem 2.9 Let A be a nonempty subset of $\mathbb{R}$ bounded above. Set
$$B = \{-a;\ a \in A\}.$$
Show that B is bounded below and
$$\inf B = -\sup A.$$

Problem 2.10 Let S and T be nonempty bounded subsets of $\mathbb{R}$ with $S \subset T$. Prove that
$$\inf T \leq \inf S \leq \sup S \leq \sup T.$$

Problem 2.11 Let $x \in \mathbb{R}$ be positive, i.e., $x \geq 0$. Show that there exists $a \in \mathbb{R}$ such that $a^2 = x$.

Problem 2.12 Let x and y be two real numbers such that $x < y$. Show that there exists a rational number r such that $x < r < y$. (In this case we say $\mathbb{Q}$ is dense in $\mathbb{R}$.) Use this result to conclude that any open nonempty interval (a, b) contains infinitely many rationals.

Problem 2.13 Let x and y be two positive real numbers such that $x < y$. Show that there exists a rational number r such that $x < r^2 < y$, without using the square-root function.

Problem 2.14 Let $\omega \in \mathbb{R}$ be an irrational positive number. Set
$$A = \{m + n\omega : \ m + n\omega > 0 \text{ and } \ m, n \in \mathbb{Z}\}.$$
Show that $\inf A = 0$.

Problem 2.15 Show that the Cantor set
$$C = \left\{\{e_n\}; e_n = 0 \text{ or } 1\right\} = \{0,1\} \times \{0,1\} \times \cdots$$
is uncountable.

Problem 2.16 If $x \geq 0$ and $y \geq 0$, show that
$$\sqrt{xy} \leq \frac{x+y}{2}.$$
When do we have equality?

Problem 2.17 Let x, y, a, and b be positive real numbers not equal to 0. Assume that $\dfrac{x}{y} < \dfrac{a}{b}$. Show that
$$\frac{x}{y} < \frac{x+a}{y+b} < \frac{a}{b}.$$

Problem 2.18 Let x and y be two real numbers. Show that
$$\frac{|x+y|}{1+|x+y|} \leq \frac{|x|}{1+|x|} + \frac{|y|}{1+|y|}.$$

CHAPTER 2. REAL NUMBERS

Problem 2.19 Let $r \in \mathbb{Q} \cap (0,1)$. Write $r = \dfrac{a}{b}$ where $a \geq 1$ and $b \geq 1$ are coprime natural numbers. Show that there exists a natural number $n \geq 1$ such that
$$\frac{1}{n+1} \leq \frac{a}{b} < \frac{1}{n}.$$
Use this to show that there exist natural numbers $n_1, \ldots, n_k$ such that
$$r = \frac{a}{b} = \frac{1}{n_1} + \cdots + \frac{1}{n_k}.$$

Problem 2.20 Let x and y be two different real numbers. Show that there exist a neighborhood X of x and a neighborhood Y of y such that $X \cap Y = \emptyset$.

Problem 2.21 Show that (a,b) is a neighborhood of any point $x \in (a,b)$.

Problem 2.22 (Young Inequality) Prove that for $p \in (1, \infty)$, we have $xy \leq \frac{1}{p}x^p + \frac{1}{q}y^q$ for $x, y \in \mathbb{R}^+ := \{x \in \mathbb{R} : x \geq 0\}$, where $q := \frac{p}{p-1}$ is the Hölder conjugate of p determined by $\frac{1}{p} + \frac{1}{q} = 1$.

Problem 2.23 (Arithmetic and Geometric Means) Prove that for $n \in \mathbb{N} \setminus \{0\}$ and $x_j \in \mathbb{R}^+$ for $1 \leq j \leq n$, one has that
$$\sqrt[n]{\prod_{j=1}^{n} x_j} \leq \frac{1}{n} \sum_{j=1}^{n} x_j.$$

Problem 2.24 (Hölder Inequality) For $p \in (1, \infty)$ and $x = (x_1, x_2, \ldots, x_n) \in \mathbb{R}^n$, define
$$|x|_p := \left(\sum_{j=1}^{n} |x_i|^p \right)^{\frac{1}{p}}.$$
Show that
$$\sum_{j=1}^{n} |x_j y_j| \leq |x|_p |y|_q \quad \text{for } x, y \in \mathbb{R}^n.$$
Note that in the case of $p = q = 2$, this reduces to the Cauchy–Schwartz Inequality.

Problem 2.25 (Minkowski Inequality) Show that for all $p \in (1, \infty)$, one has $|x+y|_p \leq |x|_p + |y|_p$ where $x, y \in \mathbb{R}^n$.

Problem 2.26 (The Nested Intersection Property) Let $\{I_n\}$ be a decreasing sequence of nonempty closed intervals in $\mathbb{R}$, i.e., $I_{n+1} \subset I_n$ for all $n \geq 1$. Show that $\bigcap_{n \geq 1} I_n$ is a nonempty closed interval. When is this intersection a single point?

Problem 2.27 (The Interval Intersection Property) Let $\{I_\alpha\}_{\alpha \in \Gamma}$ be a family of nonempty closed intervals in $\mathbb{R}$, such that $I_\alpha \cap I_\beta \neq \emptyset$ for any $\alpha, \beta \in \Gamma$. Show that $\bigcap_{\alpha \in \Gamma} I_\alpha$ is a nonempty closed interval.

Solutions

> **Solution 2.1**

Assume not. Let $r \in \mathbb{Q}$ such that $r = \sqrt{2}$ or $r^2 = 2$. Without loss of generality, we may assume $r \geq 0$. And since $r^2 = 2$, we have $r > 0$. Since r is rational, there exist two natural numbers $n \geq 1$ and $m \geq 1$ such that
$$r = \frac{n}{m}.$$
Moreover one may assume that n and m are relatively prime, i.e., the only common divisor is 1. Since $r^2 = 2$, we get
$$\left(\frac{n}{m}\right)^2 = \frac{n^2}{m^2} = 2$$
which implies $2m^2 = n^2$. Therefore, n^2 is even, so it is a multiple of 2. Assume that n is not even, then $n = 2k+1$ for some natural number k. Hence $n^2 = 4k^2 + 4k + 1 = 2(2k^2 + 2k) + 1$. In other words, if n is not even, then n^2 will not be even. Therefore, n is also even. Set $n = 2k$ for some natural number k. Then $n^2 = 4k^2$ and since $n^2 = 2m^2$, we deduce that $m^2 = 2k^2$. The same previous argument will imply that m is also even. So both n and m are even so both are multiples of 2. This is a contradiction with our assumption that both are relatively prime. Therefore, such a rational number r does not exist which completes the proof of our statement.

> **Solution 2.2**

This is an if and only if statement, and we need to prove the implications in both directions.
($\Rightarrow$): If $x = y$, then $|x - y| = 0$ and thus $|x - y| < \varepsilon$ no matter what $\varepsilon > 0$ is chosen.
:($\Leftarrow$) We give a proof by contradiction. Assume $x \neq y$, then $\varepsilon_0 = |x - y| > 0$. However, the statements
$$|x - y| = \varepsilon_0 \text{ and } |x - y| < \varepsilon_0$$
cannot be both true, our assumption is wrong, thus $x = y$.

> **Solution 2.3**

First note that
$$1 = \frac{1(1+1)}{2}$$
which implies that the desired identity holds for $n = 1$. Assume that it holds for n, and let us prove that it also holds for $n + 1$. We have
$$1 + 2 + \cdots + (n+1) = 1 + 2 + \cdots + n + (n+1).$$
Using our assumption we get
$$1 + 2 + \cdots + (n+1) = \frac{n(n+1)}{2} + (n+1).$$
Since
$$\frac{n(n+1)}{2} + (n+1) = \frac{n(n+1) + 2(n+1)}{2} = \frac{(n+1)(n+2)}{2},$$

we conclude that the identity is also valid for $n+1$. By induction we clearly showed that the above identity is valid for any natural number $n \geq 1$.

Solution 2.4

Since
$$1^2 = \frac{2+3+1}{6},$$
the above identity holds in the case $n = 1$. Assume that the identity holds for n and let us prove that it also holds for $n + 1$. Since
$$1^2 + 2^2 + \cdots + (n+1)^2 = 1^2 + 2^2 + \cdots + n^2 + (n+1)^2,$$
the induction assumption implies
$$1^2 + 2^2 + \cdots + (n+1)^2 = \frac{2n^3 + 3n^2 + n}{6} + (n+1)^2.$$

Algebraic manipulations imply
$$\frac{2n^3 + 3n^2 + n}{6} + (n+1)^2 = \frac{2n^3 + 3n^2 + n + 6(n+1)^2}{6}$$
$$= \frac{2n^3 + 3n^2 + n + 3(n+1)^2 + 3(n+1)^2}{6}$$
$$= \frac{2n^3 + 6n^2 + 7n + 3 + 3(n+1)^2}{6}.$$

On the other hand, we have
$$\frac{2(n+1)^3 + 3(n+1)^2 + (n+1)}{6} = \frac{2n^3 + 6n^2 + 6n + 2 + 3(n+1)^2 + (n+1)}{6}$$
$$= \frac{2n^3 + 6n^2 + 7n + 3 + 3(n+1)^2}{6}$$

which implies
$$1^2 + 2^2 + \cdots + (n+1)^2 = \frac{2(n+1)^3 + 3(n+1)^2 + (n+1)}{6}.$$
So our identity is also valid for $n+1$. By induction we clearly showed that the identity is valid for any natural number $n \geq 1$.

Solution 2.5

First take $n = 1$. Then $n^3 + 5n = 6$ which is a multiple of 6. Assume that $n^3 + 5n$ is divisible by 6 and let us prove that $(n+1)^3 + 5(n+1)$ is divisible by 6. But
$$(n+1)^3 + 5(n+1) = n^3 + 3n^2 + 3n + 1 + 5n + 5 = n^3 + 5n + 3(n^2 + n) + 6.$$

Next note that $n^2 + n$ is always even or a multiple of 2. Indeed if n is even, then n^2 is also even and therefore $n^2 + n$ is even. Now assume n is odd, then n^2 is also odd. Since the sum of two odd

CHAPTER 2. REAL NUMBERS

numbers is even we get that $n^2 + n$ is even. Hence $3(n^2 + n)$ is a multiple of 6. Our induction assumption implies that $n^3 + 5n$ is a multiple of 6. So $n^3 + 5n + 3(n^2 + n) + 6$ is a multiple of 6 which implies $(n+1)^3 + 5(n+1)$ is a multiple of 6. This completes our proof by induction, i.e., $n^3 + 5n$ is divisible by 6 (or multiple of 6) for all natural numbers $n \geq 1$.

Solution 2.6

It is clear that for $n = 0$, both sides of the inequality are equal to 1. Now assume that we have $(1 + x)^n \geq 1 + nx$ and let us prove that $(1 + x)^{n+1} \geq 1 + (n+1)x$. We have

$$(1 + x)^{n+1} = (1 + x)^n (1 + x) \geq (1 + nx)(1 + x)$$

and $(1 + x)^n \geq 1 + nx$. Since $(1 + nx)(1 + x) = 1 + nx + x + nx^2 = 1 + (n+1)x + nx^2$ and $nx^2 \geq 0$, we get

$$(1 + x)^{n+1} \geq 1 + (n+1)x + nx^2 \geq 1 + (n+1)x.$$

Hence the inequality is also true for $n + 1$. Therefore, by induction we have $(1 + x)^n \geq 1 + nx$ for all natural numbers $n \geq 0$.

Solution 2.7

The classical induction argument will not work here. The main reason is that in order to reach F_{n+2} one will need to make assumptions about F_{n+1} and F_n. Therefore, we will use a strong induction argument. Indeed, first it is obvious that

$$F_1 = F_2 = \frac{(1+\sqrt{5})^1 - (1-\sqrt{5})^1}{2^1 \sqrt{5}}.$$

Next assume that

$$F_k = \frac{(1+\sqrt{5})^k - (1-\sqrt{5})^k}{2^k \sqrt{5}}, k = 1, 2, \ldots, n$$

and let us prove that

$$F_{n+1} = \frac{(1+\sqrt{5})^{n+1} - (1-\sqrt{5})^{n+1}}{2^{n+1} \sqrt{5}}.$$

By the definition of the Fibonacci numbers, we have

$$F_{n+1} = F_n + F_{n-1}.$$

Our induction assumption implies

$$F_{n+1} = \frac{(1+\sqrt{5})^n - (1-\sqrt{5})^n}{2^n \sqrt{5}} + \frac{(1+\sqrt{5})^{n-1} - (1-\sqrt{5})^{n-1}}{2^{n-1} \sqrt{5}}.$$

Algebraic manipulations will imply

$$F_{n+1} = \frac{(1+\sqrt{5})^n - (1-\sqrt{5})^n + 2(1+\sqrt{5})^{n-1} - 2(1-\sqrt{5})^{n-1}}{2^n \sqrt{5}}$$

or

$$F_{n+1} = \frac{2(1+\sqrt{5})^n + 4(1+\sqrt{5})^{n-1} - 2(1-\sqrt{5})^n - 4(1-\sqrt{5})^{n-1}}{2^{n+1} \sqrt{5}}.$$

Note that
$$\begin{cases} (1+\sqrt{5})^2 = 6+2\sqrt{5} = 2(1+\sqrt{5})+4 \\ (1-\sqrt{5})^2 = 6-2\sqrt{5} = 2(1-\sqrt{5})+4 \end{cases}.$$

This easily implies
$$\begin{cases} (1+\sqrt{5})^{n+1} = 2(1+\sqrt{5})^n + 4(1+\sqrt{5})^{n-1} \\ (1-\sqrt{5})^{n+1} = 2(1-\sqrt{5})^n + 4(1-\sqrt{5})^{n-1} \end{cases}.$$

From the above equations we get
$$F_{n+1} = \frac{(1+\sqrt{5})^{n+1} - (1-\sqrt{5})^{n+1}}{2^{n+1}\sqrt{5}}.$$

The induction argument then concludes that
$$F_n = \frac{(1+\sqrt{5})^n - (1-\sqrt{5})^n}{2^n\sqrt{5}}, \; n = 1, 2, \ldots.$$

Note that the number
$$\Phi = \lim_{n \to \infty} \frac{F_{n+1}}{F_n} = \frac{1+\sqrt{5}}{2}$$
is known as the golden ratio and is one of the roots of the quadratic equation $x^2 = x + 1$.

Solution 2.8

Note that when $n = 0$, the set X is the empty set. In this case we have
$$\mathcal{P}(X) = \{\emptyset\}$$
with one element. Since $2^0 = 1$, the statement is true when $n = 0$. Assume that whenever a set X has n elements, then $\mathcal{P}(X)$ has 2^n elements. Now let us prove that whenever a set X has $n+1$ elements, then $\mathcal{P}(X)$ has 2^{n+1} elements. Indeed, let X be a set with $n+1$ elements. Fix $a \in X$ and set $Y = X \setminus \{a\}$. Then Y has n elements. Clearly, we have
$$\mathcal{P}(X) = \mathcal{P}(Y) \cup \mathcal{P}_a(Y)$$
where
$$\mathcal{P}_a(Y) = \{M \cup \{a\}; \; M \in \mathcal{P}(Y)\}.$$
The map $T_a : \mathcal{P}(Y) \to \mathcal{P}_a(Y)$ defined by
$$T(M) = M \cup \{a\}$$
is a bijection. Hence $\mathcal{P}(Y)$ and $\mathcal{P}_a(Y)$ have the same number of elements. Since Y has n elements, our assumption implies that $\mathcal{P}(Y)$ has 2^n elements. Note that $\mathcal{P}(Y)$ and $\mathcal{P}_a(Y)$ have no common point, i.e.,
$$\mathcal{P}(Y) \cap \mathcal{P}_a(Y) = \emptyset,$$
so
$$\text{number of elements of } \mathcal{P}(X) = \text{number of elements of } \mathcal{P}(Y) + \text{number of elements of } \mathcal{P}_a(Y)$$
or
$$\text{number of elements of } \mathcal{P}(X) = 2^n + 2^n = 2 \cdot 2^n = 2^{n+1}.$$

This proves our claim. So by induction we conclude that whenever a set X has n elements, then $\mathcal{P}(X)$ has 2^n elements, for any natural number $n \geq 0$.

Solution 2.9

Since A is bounded above, there exists $m \in \mathbb{R}$ such that
$$\forall a \in A \quad a \leq m.$$
Hence
$$\forall a \in A \quad -m \leq -a$$
which implies
$$\forall b \in B \quad -m \leq b.$$
So B is bounded below. Therefore $\inf B$ exists. Let us now complete the proof by showing that $\inf B = -\sup A$. By definition of $\sup A$, we know that
$$\forall a \in A \quad a \leq \sup A.$$
So
$$\forall a \in A \quad -\sup A \leq -a$$
or
$$\forall b \in B \quad -\sup A \leq b.$$
The definition of $\inf B$ implies $-\sup A \leq \inf B$. Next we have
$$\forall b \in B \quad \inf B \leq b$$
which implies
$$\forall b \in B \quad -b \leq -\inf B$$
or
$$\forall a \in A \quad a \leq -\inf B$$
since $A = \{-b;\ b \in B\}$. By the definition of $\sup A$ we get $\sup A \leq -\inf B$. Combining this conclusion with $-\sup A \leq \inf B$, we deduce that
$$\inf B = -\sup A.$$

Note that a similar proof will show that if A is bounded below, then B, defined as above, will be bounded above. Moreover we will have
$$\sup B = -\inf A.$$

In fact the above proof may be generalized to get the following result: if we set $k \cdot A = \{k \cdot a;\ a \in A\}$, for any $k \in \mathbb{R}$, then
$$\begin{cases} \sup(k \cdot A) = k \sup A & \text{provided } k \geq 0, \\ \inf(k \cdot A) = k \inf A & \text{provided } k \geq 0, \\ \sup(k \cdot A) = -k \inf A & \text{provided } k \leq 0, \\ \inf(k \cdot A) = -k \sup A & \text{provided } k \leq 0. \end{cases}$$

Solution 2.10

It is always the case that $\inf S \leq \sup S$ for any bounded nonempty subset of $\mathbb{R}$. So we need only to prove that $\inf T \leq \inf S$ and $\sup S \leq \sup T$. Let $x \in S$. Then $x \in T$ since $S \subset T$. So $\inf T \leq x$ by definition of $\inf T$. This implies that $\inf T$ is a lower bound for S because x was taken arbitrary in S. Since $\inf S$ is the greatest lower bound we get

$$\inf T \leq \inf S \ .$$

Similarly one can easily show that $\sup S \leq \sup T$.

Solution 2.11

Without loss of generality assume $x > 0$. Set

$$A = \{a \in \mathbb{R};\ a^2 \leq x\} \ .$$

Obviously we have $0 \in A$ which means that A is not empty. Next note that A is bounded above. Indeed, let $n \geq 1$ be a natural number such that $x \leq n$. We claim that n is an upper bound of A. Indeed let $a \in A$ and assume $n < a$. In particular, we have $0 < a$ which implies $n^2 < a^2$. Since $a \in A$ then we have $a^2 \leq x$ which implies $n^2 < x$. But $n \leq n^2$ which implies $n < x$, contradiction. So we must have $a \leq n$ for any $a \in A$. Since A is bounded above, then $\sup A$ exists. Set $y = \sup A$. Let us prove that $y^2 = x$ which will complete the proof of our problem. Since $0 \in A$, we get $y \geq 0$. Assume that $y^2 < x$. So the real number $\dfrac{2y+1}{x-y^2}$ is well defined. Let $n \geq 1$ be a natural number such that

$$\frac{2y+1}{x-y^2} \leq n \ ,$$

which implies

$$\frac{2y+1}{n} \leq x - y^2 \ ,$$

or

$$\frac{2y+1}{n} + y^2 \leq x \ .$$

Since $n \geq 1$ we know that $\dfrac{1}{n^2} \leq \dfrac{1}{n}$ which implies

$$y^2 + \frac{2y}{n} + \frac{1}{n^2} \leq y^2 + \frac{2y+1}{n} \leq x \ ,$$

or

$$\left(y + \frac{1}{n}\right)^2 \leq x \ .$$

Hence $y + \dfrac{1}{n} \in A$, contradicting the fact that y is an upper bound of A. So we must have $x \leq y^2$. Assume that $y^2 \neq x$. So we have $x < y^2$. In particular we have $y > 0$. Let $n \geq 1$ be a natural number such that

$$\frac{2y}{y^2 - x} \leq n \ .$$

Similar calculations as above will yield

$$x \leq \left(y - \frac{1}{n}\right)^2 \ .$$

Since $y = \sup A$, there exists $a \in A$ such that $y - \dfrac{1}{n} < a$. Since

$$\frac{1}{y} < \frac{2y}{y^2 - x},$$

we get $y - \dfrac{1}{n} > 0$. So $\left(y - \dfrac{1}{n}\right)^2 < a^2$ which implies $x < a^2$ contradicting the fact that $a \in A$. So we must have $y^2 = x$.

Solution 2.12

We have $y - x > 0$. Since $\mathbb{R}$ is Archimedean, there exists a positive integer $N \geq 1$ such that

$$N > \frac{2}{y - x}.$$

So $N(y - x) > 2$ or $Nx + 2 < Ny$. Again because $\mathbb{R}$ is Archimedean, there exists a unique integer n such that

$$n \leq Nx < n + 1.$$

We claim that $n + 1 \in (Nx, Ny)$. Indeed we have

$$Nx < n + 1 \leq Nx + 1 < Nx + 2 < Ny,$$

and since $Nx < n+1$, we have our conclusion. $N \geq 1$ implies that $x < \dfrac{n+1}{N} < y$. Take $r = \dfrac{n+1}{N}$ which completes the proof of the first statement. Next assume that the nonempty interval (a, b) has finitely many rationals. Then

$$r^* = \min\{r \in \mathbb{Q},\ a < r < b\}$$

exists and is in (a, b), i.e., $a < r^* < b$. Using the previous statement, we know that there exists a rational number q such that $a < q < r^*$. Obviously we have $q \in (a, b)$ contradicting the definition of r^*. So the set $\{r \in \mathbb{Q},\ a < r < b\}$ is infinite.

Solution 2.13

Set $A = \{r \in \mathbb{Q},\ 0 \leq r \text{ and } x < r^2\}$. Since A is not empty and bounded below, it has an infimum. Let $m = \inf A$. Clearly $0 \leq m$. We claim that $m^2 \leq x$. Assume not. Then $x < m^2$ which implies $m > 0$. Let

$$\varepsilon = \min\left(\frac{m^2 - x}{2m}, m\right).$$

It is clear that $\varepsilon > 0$. Then we have $2m\varepsilon \leq m^2 - x$ which implies

$$x \leq m^2 - 2m\varepsilon < m^2 - 2m\varepsilon + \varepsilon^2 = (m - \varepsilon)^2.$$

Since $m - \varepsilon < m$, there exists a rational number $r \in \mathbb{Q}$ such that $m - \varepsilon < r < m$. Note that r is positive because $\varepsilon \leq m$. This obviously implies $(m - \varepsilon)^2 < r^2$. In particular we have $x < (m - \varepsilon)^2 < r^2$. So $r \in A$ which contradicts $m = \inf A$. Hence our claim is valid, that is, $m^2 \leq x$. This implies $m^2 < y$. Let

$$\delta = \min\left(\frac{y - m^2}{2m + 1}, 1\right).$$

Note that $\delta > 0$. Since $\delta(2m+1) \leq y - m^2$, and $\delta \leq 1$, we get

$$(m+\delta)^2 = m^2 + 2\delta m + \delta^2 \leq m^2 + 2\delta m + \delta = m^2 + \delta(2m+1) \leq y \, .$$

Hence, $(m+\delta)^2 \leq y$. Using the characterization of the inf A, we know that there exists $r \in A$ such that $m < r < m + \delta$. So we have $r^2 < (m+\delta)^2 \leq y$ and since $r \in A$ we get

$$x < r^2 < y$$

which completes the proof of our statement.

Solution 2.14

Since all the elements of A are positive, inf A exists and is positive, i.e., inf $A \geq 0$. Assume that $\alpha = \inf A > 0$. Let us first note that $\alpha \in A$. Assume not, i.e., $\alpha \notin A$. Then, there exists $x \in A$ such that $\alpha \leq x < 2\alpha$. Since $\alpha \notin A$, then we have $\alpha < x < 2\alpha$. Again by definition of the inf A, there exists $y \in A$ such that $\alpha \leq y < x$. Using the same argument as before, we get

$$\alpha < y < x < 2\alpha \, .$$

So $x - y < 2\alpha - \alpha = \alpha$. Since $x - y \in A$, we have a contradiction with the definition of α. Therefore $\alpha \in A$. Note that $A \cap (n\alpha, (n+1)\alpha) = \emptyset$, for any natural number $n \geq 1$. Assume not, i.e., there exists a natural number $n \geq 1$ such that $A \cap (n\alpha, (n+1)\alpha) \neq \emptyset$. Let $x \in A \cap (n\alpha, (n+1)\alpha)$. Then $x - n\alpha < \alpha$ and $x - n\alpha \in A$ because of the algebraic properties satisfied by the elements of A. This will generate another contradiction with the definition of α. Therefore

$$A = \{n\alpha ; n = 1, \ldots\} \, .$$

In particular we have $\omega = n_0 \alpha$, and $1 = m_0 \alpha$ for some natural numbers $n_0 \geq 1$ and $m_0 \geq 1$. This obviously will imply

$$\frac{\omega}{1} = \frac{n_0 \alpha}{m_0 \alpha} = \frac{n_0}{m_0}$$

or that ω is rational. This is the final contradiction which forces $\alpha = 0$.

Solution 2.15

Assume not, i.e., C is countable. Hence there exists a map $f : \mathbb{N} \to C$ which is one-to-one and onto. Define $\{x_n\}$ as follows:

$$x_n = 1 - f(n)_n$$

where $f(n)_n$ is the nth element of the sequence $f(n)$ in C. Clearly $\{x_n\}$ is in C. Also note that $f(n) \neq \{x_n\}$ for any $n \in \mathbb{N}$. This contradicts the onto property of f.

Solution 2.16

We have
$$x + y - 2\sqrt{xy} = (\sqrt{x} - \sqrt{y})^2 .$$

Hence
$$x + y - 2\sqrt{xy} \geq 0$$

CHAPTER 2. REAL NUMBERS

which obviously implies $x + y \geq 2\sqrt{xy}$ or the desired inequality. Moreover we have

$$\sqrt{xy} = \frac{x+y}{2}$$

if and only if $x + y = 2\sqrt{xy}$ or $(\sqrt{x} - \sqrt{y})^2 = 0$. This will imply $\sqrt{x} = \sqrt{y}$ or $x = y$. So the equality holds if and only if $x = y$.

Solution 2.17

First, let us prove the inequality

$$\frac{x}{y} < \frac{x+a}{y+b}.$$

We have $x(y + b) - y(x + a) = xb - ya = y(\frac{x}{y}b - a) = yb(\frac{x}{y} - \frac{a}{b}) < 0$ because of the assumed inequality and all the numbers involved are positive. So $x(y + b) < y(x + a)$ or

$$\frac{x}{y} < \frac{x+a}{y+b}.$$

The other inequality follows from the same ideas.

Solution 2.18

Note that the fractions involved are always defined since the denominator cannot be 0. First assume $|x + y| \leq |x|$.

Now, since $|x+y| \leq |x|$, we get $|x+y|+|x||x+y| \leq |x|+|x||x+y|$. So $|x+y|(1+|x|) \leq |x|(1+|x+y|)$ which obviously implies

$$\frac{|x+y|}{1+|x+y|} \leq \frac{|x|}{1+|x|}.$$

Since $\dfrac{|y|}{1+|y|}$ is positive, we get

$$\frac{|x+y|}{1+|x+y|} \leq \frac{|x|}{1+|x|} + \frac{|y|}{1+|y|}.$$

If $|x + y| \leq |y|$, a similar proof will give us the above inequality. Now let us assume $\max\{|x|, |y|\} \leq |x + y|$. The triangle inequality gives $|x + y| \leq |x| + |y|$. So

$$\frac{|x+y|}{1+|x+y|} \leq \frac{|x|}{1+|x+y|} + \frac{|y|}{1+|x+y|}.$$

Since $\max\{|x|, |y|\} \leq |x + y|$, then we have

$$\frac{|x|}{1+|x+y|} + \frac{|y|}{1+|x+y|} \leq \frac{|x|}{1+|x|} + \frac{|y|}{1+|y|}.$$

Hence

$$\frac{|x+y|}{1+|x+y|} \leq \frac{|x|}{1+|x|} + \frac{|y|}{1+|y|},$$

which completes the proof of the desired inequality.

Solution 2.19

Using the Archimedean property of the reals, we know that there exists a unique integer m such that
$$m \leq \frac{b}{a} < m+1 \ .$$
Since $\frac{b}{a} > 1$, we get $m \geq 1$. Easy algebra manipulations give
$$\frac{1}{m+1} < \frac{a}{b} \leq \frac{1}{m} \ .$$
If $\frac{b}{a} = \frac{1}{m}$, then we must have $m > 1$. In this case take $n = m-1$. Otherwise take $n = m$ to get
$$\frac{1}{n+1} \leq \frac{a}{b} < \frac{1}{n} \ .$$
To prove the second part we will use the strong induction argument. If $a = 1$, then the conclusion is obvious. Assume the conclusion is true for $a = 1, \ldots, k$ and let us prove that the conclusion is also true for $a = k+1$. Let $b \geq 1$ be a natural number coprime with $k+1$ such that $k+1 < b$. Then, the first part implies the existence of a natural number $n \geq 1$ such that
$$\frac{1}{n+1} \leq \frac{k+1}{b} < \frac{1}{n} \ .$$
If $\frac{1}{n+1} = \frac{k+1}{b}$, then we have nothing to prove. Otherwise assume $\frac{1}{n+1} < \frac{k+1}{b}$. Then
$$0 < \frac{k+1}{b} - \frac{1}{n+1} < \frac{1}{n} - \frac{1}{n+1} < 1 \ .$$
Set
$$r^* = \frac{k+1}{b} - \frac{1}{n+1} = \frac{a^*}{b^*} \in \mathbb{Q} \ ,$$
where a^* and b^* are natural coprime numbers. Obviously we have $a^* \leq (k+1)(n+1) - b$. Since $\frac{(k+1)}{b} < \frac{1}{n}$, then $(k+1)n < b$ or $(k+1)(n+1) - b < k+1$. This implies $a^* < k+1$. Our induction assumption implies
$$\frac{a^*}{b^*} = \frac{1}{n_1} + \cdots + \frac{1}{n_l}$$
for some natural numbers $n_1, \ldots, n_l$. Hence
$$\frac{k+1}{b} = \frac{1}{n+1} + \frac{1}{n_1} + \cdots + \frac{1}{n_l} \ .$$
Therefore, by induction we have proved our initial claim.

Solution 2.20

Since $x \neq y$, then we have
$$\varepsilon = \frac{|x-y|}{3} > 0 \ .$$

CHAPTER 2. REAL NUMBERS

Set
$$X = (x-\varepsilon, x+\varepsilon) \text{ and } Y = (y-\varepsilon, y+\varepsilon).$$

Let us check that $X \cap Y = \emptyset$. Assume not. And let $z \in X \cap Y$. So
$$|z-x| < \varepsilon \text{ and } |z-y| < \varepsilon.$$

Hence
$$|x-y| = |x-z+z-y| \leq |x-z| + |z-y| < \varepsilon + \varepsilon = \frac{2|x-y|}{3} < |x-y|.$$

This contradiction proves our claim.

Solution 2.21

Let $x \in (a,b)$. Since $x \neq a$ and $x \neq b$, we have
$$\varepsilon = \frac{1}{3}\min(|x-a|, |b-x|) > 0.$$

Let us show that $(x-\varepsilon, x+\varepsilon) \subset (a,b)$. Indeed, we have
$$x + \varepsilon = x + \frac{1}{3}\min(|x-a|, |b-x|) < x + |x-b| = x + b - x = b$$

because $x < b$. Similarly we have
$$x - \varepsilon = x - \frac{1}{3}\min(|x-a|, |b-x|) > x - |x-a| = x - (x-a) = a.$$

These two inequalities obviously imply
$$(x-\varepsilon, x+\varepsilon) \subset (a,b),$$

which implies that (a,b) is a neighborhood of any point $x \in (a,b)$.

Solution 2.22

Suppose $x, y \in (0, \infty)$. From the concavity of the logarithim function and the fact that $\frac{1}{p} + \frac{1}{q} = 1$, this implies that $\log\left(\frac{x^p}{p} + \frac{y^q}{q}\right) \geq \frac{1}{p}\log x^p + \frac{1}{q}\log y^q = \log x + \log y = \log xy$. Since the exponential function is increasing and $e^{\log x} = x$ for all x, the claimed inequality follows.

Solution 2.23

We use induction to prove. For $n=1$, the above inequality is true. Suppose it is true for some $n \geq 1$. Then
$$\sqrt[n+1]{\prod_{j=1}^{n+1} x_j} \leq \left(\frac{1}{n}\sum_{j=1}^{n} x_j\right)^{\frac{n}{n+1}} (x_{n+1})^{\frac{1}{n+1}}.$$

To the right-hand side of this inequality, apply the Young Inequality (previous problem) with
$$x = \left(\frac{1}{n}\sum_{j=1}^{n} x_j\right)^{\frac{n}{n+1}}, \ y = (x_{n+1})^{\frac{1}{n+1}}, \text{ and } p = 1 + \frac{1}{n}. \text{ Then}$$

$$xy \le \frac{1}{p}x^p + \frac{1}{q}y^q = \frac{1}{n+1}\sum_{j=1}^{n}x_j + \frac{1}{n+1}(x_{n+1}) = \frac{1}{n+1}\sum_{j=1}^{n+1}x_j.$$

Solution 2.24

Without loss of generality, take $x \ne 0$ and $y \ne 0$. From the Young Inequality,

$$\frac{|x_j|}{|x|_p} \cdot \frac{|y_j|}{|y|_q} \le \frac{1}{p} \cdot \frac{|x_j|^p}{|x|_p^p} + \frac{1}{q} \cdot \frac{|y_j|^q}{|y|_q^q}, \quad \text{for } 1 \le j \le n.$$

Summing this inequality over j yields

$$\frac{\sum_{j=1}^{n}|x_j y_j|}{|x|_p |y|_q} \le \frac{1}{p} + \frac{1}{q} = 1$$

and therefore the desired inequality follows.

Solution 2.25

Applying the triangle inequality, we get

$$|x+y|_p^p = \sum_{j=1}^{n}|x_j+y_j|^{p-1}\cdot|x_j+y_j| \le \sum_{j=1}^{n}|x_j+y_j|^{p-1}\cdot|x_j| + \sum_{j=1}^{n}|x_j+y_j|^{p-1}\cdot|y_j|.$$

Thus the Hölder Inequality implies

$$|x+y|_p^p \le |x|_p\left(\sum_{j=1}^{n}|x_j+y_j|^p\right)^{\frac{1}{q}} + |y|_p\left(\sum_{j=1}^{n}|x_j+y_j|^p\right)^{\frac{1}{q}} = (|x|_p+|y|_p)|x+y|_p^{\frac{p}{q}}.$$

If $x+y = 0$, then the inequality is true. Otherwise, divide both sides of the inequality by $|x+y|_p^{\frac{p}{q}}$ to get $|x+y|_p^{p-\frac{p}{q}} \le |x|_p + |y|_p$. Since $p - \frac{p}{q} = 1$, the claim follows.

Solution 2.26

Set $I_n = [a_n, b_n]$. Then our assumption on $\{I_n\}$ implies

$$a_n \le a_{n+1} \le b_{n+1} \le b_n$$

for all $n \ge 1$. Since $I_n \subset I_1$, both $\{a_n\}$ and $\{b_n\}$ are bounded. Set

$$a_\infty = \sup\{a_n; n \ge 1\} \text{ and } b_\infty = \inf\{b_n; n \ge 1\}.$$

From the inequalities $a_n \le a_{n+1} \le b_{n+1} \le b_n$, for all $n \ge 1$, we get

$$a_n \le a_\infty \le b_\infty \le b_n$$

for all $n \ge 1$. Indeed, for any $n, m \ge 1$, we have $a_n \le b_m$. This follows from the fact that $\{a_n\}$ is increasing and $\{b_n\}$ is decreasing. By definition of the supremum and infimum, we get $a_\infty \le b_\infty$.

The inequalities $a_n \leq a_\infty$ and $b_\infty \leq b_n$ follow easily. Next let us complete the proof by showing that
$$\bigcap_{n \geq 1} I_n = [a_\infty, b_\infty] \,.$$
From the above inequalities we have $[a_\infty, b_\infty] \subset [a_n, b_n] = I_n$ for all $n \geq 1$. Hence we are only left to prove $\bigcap_{n \geq 1} I_n \subseteq [a_\infty, b_\infty]$. So let $x \in I_n$ for all $n \geq 1$. Let us prove that $x \in [a_\infty, b_\infty]$. Since $x \in I_n$, we get $a_n \leq x \leq b_n$ for all $n \geq 1$. By definition of the supremum and infimum, we get $a_\infty \leq x \leq b_\infty$, or $x \in [a_\infty, b_\infty]$. Clearly the intersection is a single point if $a_\infty = b_\infty$. This happens if and only if $b_n - a_n$ goes to 0 as n goes to ∞.

Solution 2.27

Let us set $I_\alpha = [a_\alpha, b_\alpha]$. First let us show that the condition satisfied by the intervals implies $a_\alpha \leq b_\beta$ for any $\alpha, \beta \in \Gamma$. Indeed, fix $\alpha, \beta \in \Gamma$, then consider $x \in I_\alpha \cap I_\beta$ because $I_\alpha \cap I_\beta \neq \emptyset$. Hence $a_\alpha \leq x \leq b_\beta$ which implies the desired inequality. This will force the sets $\{a_\alpha\}$ and $\{b_\alpha\}$ to be bounded. Set
$$a_\infty = \sup\{a_\alpha; \alpha \in \Gamma\} \quad \text{and} \quad b_\infty = \inf\{b_\alpha; \alpha \in \Gamma\} \,.$$
Following the same argument in the previous problem one will easily check that $\bigcap_{\alpha \in \Gamma} I_\alpha = [a_\infty, b_\infty]$.

Chapter 3

Sequences

Mathematicians have tried in vain to this day to discover some order in the sequence of prime numbers, and we have reason to believe that it is a mystery into which the human mind will never penetrate.

Leonhard Euler (1707–1783)

- A *sequence* is a function whose domain is the set $\mathbb{N}$ of natural numbers.

- A sequence $\{x_n\}$ is said to *converge* to a real number x, provided that for each $\varepsilon > 0$ there exists an integer N such that $n \geq N$ implies that $|x_n - x| < \varepsilon$.
  In this case we also say that $\{x_n\}$ converges to x, or x is the limit of $\{x_n\}$, and we write $x_n \to x$, or $\lim_{n \Rightarrow \infty} x_n = x$. If $\{x_n\}$ does not converge, it is said to *diverge*.

- A sequence $\{x_n\}$ is said to be *bounded* if the range $\{x_n : n \in \mathbb{N}\}$ is a bounded set, that is, if there exists $M \geq 0$ such that $|x_n| \leq M$ for all $n \in \mathbb{N}$.

- *Bolzano–Weierstrass Theorem*: Every bounded sequence has a convergent subsequence.

- Let $\{x_n\}_{n=1}^{\infty}$ be a sequence and for each $n \in \mathbb{N}$, set
$$y_n = \sup\{x_k : k \geq n\}.$$
The *limit superior* of $\{x_n\}$, denoted by $\limsup\{x_n\}$ or $\overline{\lim}\{x_n\}$, is defined by
$$\overline{\lim}\{x_n\} = \inf\{y_n : n \in \mathbb{N}\} = \inf\{x : x = \sup\{x_k : k \geq n\} \text{ for some } n \in \mathbb{N}\}$$

provided that the quantity on the right exists. Likewise we define the *limit inferior* by

$$\underline{\lim}\{x_n\} = \sup\{x : x = \inf\{x_k : k \geq n\} \text{ for some } n \in \mathbb{N}\}.$$

It is well known that if $\{x_n\}$ is a sequence, then $\{x_n\}$ has a limit if and only if the limit superior and the limit inferior exist and are equal.

- A sequence $\{x_n\}$ of real numbers is said to be a *Cauchy sequence* if for every $\varepsilon > 0$, there is an integer N such that
$$|x_n - x_m| < \varepsilon \text{ if } n \geq N \text{ and } m \geq N.$$

- Let $\{x_n\}_{n=1}^\infty$ be a sequence and let $\{n_k\}_{k=1}^\infty$ be any sequence of natural numbers such that $n_1 < n_2 < n_3 < \ldots$. The sequence $\{x_{n_k}\}_{k=1}^\infty$ is called a *subsequence* of $\{x_n\}_{n=1}^\infty$.

Problem 3.1 Show that each bounded sequence of real numbers has a convergent subsequence.

Problem 3.2 Show that if $\{x_n\}$ converges to l, then $\{|x_n|\}$ converges to $|l|$. What about the converse?

Problem 3.3 Let C be a real number such that $|C| < 1$. Show that $\lim_{n \to \infty} C^n = 0$.

Problem 3.4 Let $\{x_n\}$ be a sequence such that $\{x_{2n}\}$, $\{x_{2n+1}\}$, and $\{x_{3n}\}$ are convergent. Show that $\{x_n\}$ is convergent.

Problem 3.5 Let S be a nonempty subset of $\mathbb{R}$ which is bounded above. Set $s = \sup S$. Show that there exists a sequence $\{x_n\}$ in S which converges to s.

Problem 3.6 Let $\{x_n\}$ and $\{y_n\}$ be two real sequences such that

(a) $x_n \leq y_n$ for all n;

(b) $\{x_n\}$ is increasing;

(c) $\{y_n\}$ is decreasing.

Show that $\{x_n\}$ and $\{y_n\}$ are convergent and

$$\lim_{n \to \infty} x_n \leq \lim_{n \to \infty} y_n.$$

When do we have equality of the limits?

CHAPTER 3. SEQUENCES

Problem 3.7 Show that $\{x_n\}$ defined by
$$x_n = 1 + \frac{1}{2} + \cdots + \frac{1}{n}$$
is divergent.

Problem 3.8 Show that $\{x_n\}$ defined by
$$x_n = 1 + \frac{1}{2} + \cdots + \frac{1}{n} - \ln(n)$$
is convergent.

Problem 3.9 Show that the sequence $\{x_n\}$ defined by
$$x_n = \int_1^n \frac{\cos(t)}{t^2} dt$$
is Cauchy.

Problem 3.10 Let $\{x_n\}$ be a sequence such that there exist $A > 0$ and $C \in (0,1)$ for which
$$|x_{n+1} - x_n| \leq AC^n$$
for any $n \geq 1$. Show that $\{x_n\}$ is Cauchy. Is this conclusion still valid if we assume only
$$\lim_{n \to \infty} |x_{n+1} - x_n| = 0?$$

Problem 3.11 Show that if a subsequence $\{x_{n_k}\}$ of a Cauchy sequence $\{x_n\}$ is convergent, then $\{x_n\}$ is convergent.

Problem 3.12 Discuss the convergence or divergence of
$$x_n = \frac{n^2}{\sqrt{n^6+1}} + \frac{n^2}{\sqrt{n^6+2}} + \cdots + \frac{n^2}{\sqrt{n^6+n}}.$$

Problem 3.13 Discuss the convergence or divergence of
$$x_n = \frac{[\alpha] + [2\alpha] + \cdots + [n\alpha]}{n^2},$$
where $[x]$ denotes the greatest integer less than or equal to the real number x, and α is an arbitrary real number.

Problem 3.14 Discuss the convergence or divergence of
$$x_n = \frac{\alpha^n - \beta^n}{\alpha^n + \beta^n}$$
where α and β are real numbers such that $|\alpha| \neq |\beta|$.

Problem 3.15 (**Cesaro Average**) Let $\{x_n\}$ be a real sequence which converges to l. Show that the sequence
$$y_n = \frac{x_1 + x_2 + \cdots + x_n}{n}$$
also converges to l. What about the converse? As an application of this, show that if $\{x_n\}$ is such that $\lim_{n \to \infty} x_{n+1} - x_n = l$, then
$$\lim_{n \to \infty} \frac{x_n}{n} = l \ .$$

Problem 3.16 Let $\{x_n\}$ be a real sequence with $x_n \neq 0$. Assume that
$$\lim_{n \to \infty} \frac{x_{n+1}}{x_n} = l \ .$$
Show that

(a) if $|l| < 1$, then $\lim_{n \to \infty} x_n = 0$;

(b) and if $|l| > 1$, then $\{x_n\}$ is divergent.

What happens when $|l| = 1$? As an application decide on convergence or divergence of
$$x_n = \frac{\alpha^n}{n^k} \quad \text{and} \quad y_n = \frac{\alpha^n}{n!} \ .$$

Problem 3.17 Given $x \geq 1$, show that
$$\lim_{n \to \infty} \left(2 \sqrt[n]{x} - 1\right)^n = x^2 \ .$$

Problem 3.18 Show that
$$\lim_{n \to \infty} \frac{\left(2 \sqrt[n]{n} - 1\right)^n}{n^2} = 1 \ .$$

CHAPTER 3. SEQUENCES

Problem 3.19 Let $\{x_n\}$ defined by
$$x_1 = 1 \text{ and } x_{n+1} = \frac{1}{2}\left(x_n + \frac{2}{x_n}\right).$$
Show that $\{x_n\}$ is convergent and find its limit.

Problem 3.20 Let $\{x_n\}$ be a sequence defined by
$$x_1 = 1, \text{ and } x_{n+1} = \sqrt{x_n^2 + \frac{1}{2^n}}.$$
Show that $\{x_n\}$ is convergent.

Problem 3.21 For any $n \in \mathbb{N}$ set $I_n = \int_0^{\pi/2} \cos^n(t)\,dt$, known as Wallis integrals.

1. Show that $(n+2)I_{n+2} = (n+1)I_n$. Then use it to find explicitly I_{2n} and I_{2n+1}.

2. Show that $\displaystyle\lim_{n\to\infty} \frac{I_{n+1}}{I_n} = 1$.

3. Show that $\{(n+1)I_n I_{n+1}\}$ is a constant sequence. Then conclude that
$$\lim_{n\to\infty} I_n\sqrt{2n} = \sqrt{\pi}.$$

Problem 3.22 Consider the sequence
$$x_n = \frac{n!}{\sqrt{n}}\left(\frac{e}{n}\right)^n, \; n = 1,\ldots.$$

1. Show that $\{\ln(x_n)\}$ is convergent. Use this to show that $\{x_n\}$ is convergent.

2. Use Wallis integrals to find the limit of $\{x_n\}$.

3. Use 1. and 2. to prove the Stirling formula
$$n! \approx \left(\frac{n}{e}\right)^n \sqrt{2\pi n}$$
when $n \to \infty$.

Problem 3.23 Find the limit superior and limit inferior of the sequence $\{x_n\}$, where

- $x_n = 1 + (-1)^n + \frac{1}{2^n}$
- $x_n = 2^n$

Problem 3.24 Let $\{x_n\}$ be a bounded sequence. Prove there exists a subsequence of $\{x_n\}$ which converges to $\liminf\limits_{n\to\infty} x_n$. Show that the same conclusion holds for $\limsup\limits_{n\to\infty} x_n$.

Problem 3.25 Let $\{x_n\}$ be a sequence and let $\{x_{n_k}\}$ be any of its subsequences. Show that

$$\liminf_{n\to\infty} x_n \le \liminf_{n_k\to\infty} x_{n_k} \le \limsup_{n_k\to\infty} x_{n_k} \le \liminf_{n\to\infty} x_n.$$

In particular, if $\{x_{n_k}\}$ is convergent, then

$$\liminf_{n\to\infty} x_n \le \lim_{n_k\to\infty} x_{n_k} \le \limsup_{n\to\infty} x_n.$$

Is the converse true? That is, for any l between $\liminf\limits_{n\to\infty} x_n$ and $\limsup\limits_{n\to\infty} x_n$, there exists a subsequence $\{x_{n_k}\}$ which converges to l.

Problem 3.26 If (x_n) and (y_n) are bounded real sequences, show that

$$\limsup_{n\to\infty}(x_n + y_n) \le \limsup_{n\to\infty} x_n + \limsup_{n\to\infty} y_n.$$

Do we have equality?

Problem 3.27 If $x_n > 0$, $n = 1, 2, \ldots$, show that

$$\liminf_{n\to\infty} \frac{x_{n+1}}{x_n} \le \liminf_{n\to\infty} \sqrt[n]{x_n} \le \limsup_{n\to\infty} \sqrt[n]{x_n} \le \limsup_{n\to\infty} \frac{x_{n+1}}{x_n}.$$

Deduce that if $\lim\limits_{n\to\infty} \dfrac{x_{n+1}}{x_n}$ exists, then $\lim_{n\to\infty} \sqrt[n]{x_n}$ exists. What happened to the converse?

CHAPTER 3. SEQUENCES

Solutions

Solution 3.1

Let $(x_n : n \in \mathbb{N})$ be a bounded sequence, say $|x_n| \leq M$ for all n.

Let $I_0 = [-M, M]$, $a_0 = -M$, and $b_0 = M$, so that $I_0 = [a_0, b_0]$ and I_0 contains infinitely many of the x_n (in fact, all of them).

We construct inductively a sequence of intervals $I_k = [a_k, b_k]$ such that I_k contains infinitely many of the x_n and $b_k - a_k = 2M/2^k$. This certainly holds for $k = 0$.

Suppose it holds for some value of k. Then at least one of the intervals $[a_k, (a_k + b_k)/2]$ and $[(a_k + b_k)/2, b_k]$ contains infinitely many of the x_n. If the former, then let $a_{k+1} = a_k$, $b_{k+1} = (a_k + b_k)/2$. Otherwise, let $a_{k+1} = (a_k + b_k)/2$, $b_{k+1} = b_k$. In either case, the interval $I_{k+1} = [a_{k+1}, b_{k+1}]$ contains infinitely many of the x_n, and

$$b_{k+1} - a_{k+1} = \frac{1}{2}(b_k - a_k) = \left(\frac{1}{2}\right)^{k+1} \times 2M.$$

This completes the inductive construction.

Clearly $a_0 \leq a_1 \leq a_2 \leq \cdots \leq b_2 \leq b_1 \leq b_0$. Thus (a_n) is an increasing bounded sequence, so by completeness has a limit, say x. Moreover since each b_k is an upper bound for (a_n) and x is the supremum, $x \leq b_k$ for each k. Thus $a_k \leq x \leq b_k$ for every k. In other words, $x \in I_k$ for every k.

We now construct inductively a subsequence (x_{n_k}) of (x_n) such that $x_{n_k} \in I_k$ for every k. Let $x_{n_0} = x_0$. Assuming x_{n_k} has been chosen, let n_{k+1} be the least $n > n_k$ such that $x_n \in I_{k+1}$. Then (x_{n_k}) is a subsequence of (x_n), and $x_{n_k} \in I_k$ for every k.

Since x_{n_k} and x both lie in the same interval I_k of length $2M/2^k$, it follows that

$$|x_{n_k} - x| \leq \left(\frac{1}{2}\right)^k \times 2M$$

and so $|x_{n_k} - x| \to 0$ as $n \to \infty$. Thus (x_{n_k}) is a convergent subsequence of (x_n), as required.

Solution 3.2

Note that for any real numbers $x, y \in \mathbb{R}$, we have

$$\Big||x| - |y|\Big| \leq |x - y|.$$

Since $\{x_n\}$ converges to l, then for any $\varepsilon > 0$, there exists $n_0 \geq 1$ such that for any $n \geq n_0$, we have

$$|x_n - l| < \varepsilon.$$

Hence

$$\Big||x_n| - |l|\Big| < \varepsilon$$

for any $n \geq n_0$. This obviously implies the desired conclusion. For the converse, take $x_n = (-1)^n$, for $n = 0, \ldots$. Then we have $|x_n| = 1$ which means that $\{|x_n|\}$ converges to 1. But $\{x_n\}$ does not converge. Note that if $l = 0$, then the converse is true.

Solution 3.3

If $C = 0$, then the conclusion is obvious. Assume first $0 < C < 1$. Then the sequence $\{C^n\}$ is decreasing and bounded below by 0. So it has a limit L. Let us prove that $L = 0$. We have $C^{n+1} = CC^n$ so by passing to the limit we get $L = CL$ which implies $L = 0$. If $-1 < -C < 0$, then we use $(-C)^n = (-1)^n C^n$ and the fact that the product of a bounded sequence with a sequence which converges to 0 also converges to 0 to get $\lim_{n \to \infty} (-C)^n = 0$. Therefore, for any $-1 < C < 1$, we have $\lim_{n \to \infty} C^n = 0$.

Solution 3.4

If $\{x_n\}$ is convergent, then all subsequences of $\{x_n\}$ are convergent and converge to the same limit. Therefore, let us show that the three subsequences converge to the same limit. Write

$$\lim_{n \to \infty} x_{2n} = \alpha_1, \ \lim_{n \to \infty} x_{2n+1} = \alpha_2, \ \text{and} \ \lim_{n \to \infty} x_{3n} = \alpha_3 \ .$$

The sequence $\{x_{6n}\}$ is a subsequence of both sequences $\{x_{2n}\}$ and $\{x_{3n}\}$. Hence $\{x_{6n}\}$ converges and forces the following:

$$\lim_{n \to \infty} x_{6n} = \lim_{n \to \infty} x_{2n} = \lim_{n \to \infty} x_{3n}$$

or $\alpha_1 = \alpha_3$. On the other hand, the sequence $\{x_{6n+3}\}$ is a subsequence of both sequences $\{x_{2n+1}\}$ and $\{x_{3n}\}$. Hence $\{x_{6n+3}\}$ converges and forces the following:

$$\lim_{n \to \infty} x_{6n+3} = \lim_{n \to \infty} x_{2n+1} = \lim_{n \to \infty} x_{3n}$$

or $\alpha_2 = \alpha_3$. Hence $\alpha_1 = \alpha_2 = \alpha_3$. Let us write

$$\lim_{n \to \infty} x_{2n} = \lim_{n \to \infty} x_{2n+1} = l$$

and let us prove that $\lim_{n \to \infty} x_n = l$. Let $\varepsilon > 0$. There exist $N_0 \geq 1$ and $N_1 \geq 1$ such that

$$\begin{cases} |x_{2n} - l| < \varepsilon & \text{for all } n \geq N_0, \\ |x_{2n+1} - l| < \varepsilon & \text{for all } n \geq N_1. \end{cases}$$

Set $N = \max\{2N_0, 2N_1 + 1\}$. Let $n \geq N$. If $n = 2k$, then we have $k \geq N_0$ since $n \geq N \geq 2N_0$. Using the above inequalities we get $|x_{2k} - l| < \varepsilon$ or $|x_n - l| < \varepsilon$. A similar argument when n is odd will yield the same inequality. Therefore

$$|x_n - l| < \varepsilon$$

for any $n \geq N$. This completes the proof of our statement.

CHAPTER 3. SEQUENCES

Solution 3.5

By the characterization of the supremum, we know that for any $\varepsilon > 0$ there exists $x \in S$ such that
$$s - \varepsilon < x \leq s \,.$$
So for any $n \geq 1$, there exists $x_n \in S$ such that
$$s - \frac{1}{n} < x_n \leq s \,.$$
Since $\left\{\dfrac{1}{n}\right\}$ goes to 0, given $\varepsilon > 0$, there exists $n_0 \geq 1$ such that for any $n \geq n_0$ we have $\dfrac{1}{n} < \varepsilon$. So for any $n \geq n_0$ we have
$$s - \varepsilon < s - \frac{1}{n} < x_n \leq s < s + \varepsilon \,,$$
which implies
$$|x_n - s| < \varepsilon \,,$$
which translates into $\lim_{n \to \infty} x_n = s$.

Solution 3.6

Since $\{y_n\}$ is decreasing, we have $y_n \leq y_1$ for $n \geq 1$. So for any $n \geq 1$ we have $x_n \leq y_n \leq y_1$. This implies that $\{x_n\}$ is bounded above. Since it is increasing it converges. Similar argument shows that $\{y_n\}$ is bounded below and therefore converges as well. From (a) we get the desired inequality on the limits. In order to have the equality of the limits we must have $\lim_{n \to \infty} y_n - x_n = 0$. This result is useful when dealing with nested intervals in $\mathbb{R}$ and alternating real series.

Solution 3.7

We have
$$x_{2n} - x_n = \frac{1}{n+1} + \frac{1}{n+2} + \cdots + \frac{1}{2n}$$
for any $n \geq 1$. So
$$\frac{1}{n+n} + \frac{1}{n+n} + \cdots + \frac{1}{2n} \leq x_{2n} - x_n$$
or $\dfrac{1}{2} \leq x_{2n} - x_n$. This clearly implies that $\{x_n\}$ fails to be Cauchy. Therefore it diverges.

Solution 3.8

Though real functions will be handled in the next chapters, here we will use the integral definition of the logarithm function. In particular, we have
$$\ln(x) = \int_1^x \frac{1}{t} dt \,.$$
In this case if $0 < a < b$, then we have
$$\frac{b-a}{b} \leq \int_a^b \frac{1}{t} dt \leq \frac{b-a}{a}.$$

Since
$$\ln(n) = \int_1^n \frac{1}{t} dt = \sum_{k=1}^{n-1} \int_k^{k+1} \frac{1}{t} dt ,$$
we get
$$\ln(n) \leq \sum_{k=1}^{n-1} \frac{k+1-k}{k} = 1 + \frac{1}{2} + \cdots + \frac{1}{n-1} .$$
Hence
$$x_n = 1 + \frac{1}{2} + \cdots + \frac{1}{n} - \ln(n) = 1 + \frac{1}{2} + \cdots + \frac{1}{n-1} - \ln(n) + \frac{1}{n} > 0 .$$
On the other hand, we have
$$x_{n+1} - x_n = \frac{1}{n+1} - \ln(n+1) + \ln(n) = \frac{1}{n+1} - \int_n^{n+1} \frac{1}{t} dt < 0 .$$

These two inequalities imply that $\{x_n\}$ is decreasing and bounded below by 0. Therefore $\{x_n\}$ is convergent. Its limit is known as the Euler constant.

$\boxed{\text{Solution 3.9}}$

For any natural integers $n < m$ we have
$$\left| \int_n^m \frac{\cos(t)}{t^2} dt \right| \leq \int_n^m \frac{|\cos(t)|}{t^2} dt \leq \int_n^m \frac{1}{t^2} dt = \left[-\frac{1}{t} \right]_n^m = \frac{1}{m} - \frac{1}{n}.$$

Since $\lim_{n\to\infty} \frac{1}{n} = 0$, then for any $\varepsilon > 0$, there exists $n_0 \geq 1$ such that for any $n \geq n_0$ we have $\frac{1}{n} < \varepsilon$. So for $n, m \geq n_0$, $n \leq m$, we have
$$|x_n - x_m| = \left| \int_n^m \frac{\cos(t)}{t^2} dt \right| \leq \frac{1}{m} - \frac{1}{n} < \varepsilon ,$$
which shows that $\{x_n\}$ is a Cauchy sequence.

$\boxed{\text{Solution 3.10}}$

Let $n \geq 1$ and $h \geq 1$. We have
$$|x_{n+h} - x_n| = \left| \sum_{k=0}^{h-1} x_{n+k+1} - x_{n+k} \right| \leq \sum_{k=0}^{h-1} |x_{n+k+1} - x_{n+k}| .$$

Our assumption on $\{x_n\}$ implies
$$|x_{n+h} - x_n| \leq \sum_{k=0}^{h-1} AC^{n+k} = AC^n \frac{1 - C^h}{1 - C} < A \frac{C^n}{1 - C} .$$

Since $0 < C < 1$, $\lim_{n\to\infty} C^n = 0$. Hence
$$\lim_{n\to\infty} A \frac{C^n}{1 - C} = 0 .$$

CHAPTER 3. SEQUENCES

This will force $\{x_n\}$ to be Cauchy. The second part of the statement is not true. Indeed, take $x_n = \sqrt{n}$. Then we have

$$\lim_{n \to \infty} \sqrt{n+1} - \sqrt{n} = \lim_{n \to \infty} \frac{1}{\sqrt{n+1} + \sqrt{n}} = 0.$$

But the sequence $\{x_n\}$ is divergent.

Solution 3.11

Set $\lim_{n_k \to \infty} x_{n_k} = L$. Let us show that $\{x_n\}$ converges to L. Let $\varepsilon > 0$. Since $\{x_n\}$ is Cauchy, there exists $n_0 \geq 1$ such that for any $n, m \geq n_0$ we have

$$|x_n - x_m| < \frac{\varepsilon}{2}.$$

Since $\lim_{n_k \to \infty} x_{n_k} = L$, there exists $k_0 \geq 1$ such that for any $k \geq k_0$ we have

$$|x_{n_k} - L| < \frac{\varepsilon}{2}.$$

For k big enough to have $n_k \geq n_0$ we get

$$|x_n - L| \leq |x_n - x_{n_k}| + |x_{n_k} - L| < \frac{\varepsilon}{2} + \frac{\varepsilon}{2} = \varepsilon$$

for any $n \geq n_0$. This completes the proof.

Solution 3.12

Note that for any $k = 1, \ldots, n$, we have

$$\frac{n^2}{\sqrt{n^6 + n}} \leq \frac{n^2}{\sqrt{n^6 + k}} \leq \frac{n^2}{\sqrt{n^6}} = \frac{1}{n}$$

which implies

$$n \frac{n^2}{\sqrt{n^6 + n}} \leq x_n \leq n \frac{1}{n}$$

or

$$\frac{n^3}{\sqrt{n^6 + n}} \leq x_n \leq 1.$$

Because

$$\frac{n^3}{\sqrt{n^6 + n}} = \frac{n^3}{n^3 \sqrt{1 + \frac{1}{n^2}}} = \frac{1}{\sqrt{1 + \frac{1}{n^2}}}$$

and $\lim_{n \to \infty} \frac{1}{n^2} = 0$, then $\lim_{n \to \infty} \frac{n^3}{\sqrt{n^6 + n}} = 1$. The Squeeze Theorem forces the conclusion

$$\lim_{n \to \infty} \frac{n^2}{\sqrt{n^6 + 1}} + \frac{n^2}{\sqrt{n^6 + 2}} + \cdots + \frac{n^2}{\sqrt{n^6 + n}} = 1.$$

Solution 3.13

By definition of the greatest integer function $[\cdot]$, we have

$$[x] \leq x < [x] + 1$$

for any real number x. This will easily imply $x - 1 < [x] \leq x$. So

$$\frac{(\alpha - 1) + (2\alpha - 1) + \cdots + (n\alpha - 1)}{n^2} < \frac{[\alpha] + [2\alpha] + \cdots + [n\alpha]}{n^2} \leq \frac{\alpha + 2\alpha + \cdots + n\alpha}{n^2}$$

or

$$\frac{(1 + 2 + \cdots + n)\alpha - n}{n^2} < \frac{[\alpha] + [2\alpha] + \cdots + [n\alpha]}{n^2} \leq \frac{(1 + 2 + \cdots + n)\alpha}{n^2}.$$

The algebraic identity $1 + 2 + \cdots + m = \dfrac{m(m+1)}{2}$ for any natural number $m \geq 1$ gives

$$\frac{\frac{n(n+1)}{2}\alpha - n}{n^2} < \frac{[\alpha] + [2\alpha] + \cdots + [n\alpha]}{n^2} \leq \frac{\frac{n(n+1)}{2}\alpha}{n^2}$$

or

$$\frac{(n+1)\alpha}{2n} - \frac{1}{n} < \frac{[\alpha] + [2\alpha] + \cdots + [n\alpha]}{n^2} \leq \frac{(n+1)\alpha}{2n}.$$

Since

$$\lim_{n \to \infty} \frac{(n+1)\alpha}{2n} - \frac{1}{n} = \frac{\alpha}{2} \quad \text{and} \quad \lim_{n \to \infty} \frac{(n+1)\alpha}{2n} = \frac{\alpha}{2},$$

the Squeeze Theorem implies $\lim_{n \to \infty} x_n = \dfrac{\alpha}{2}$.

Solution 3.14

We have two cases, either $|\alpha| < |\beta|$ or $|\alpha| > |\beta|$. Assume first that $|\alpha| < |\beta|$. Set $r = \frac{\alpha}{\beta}$. Then algebraic manipulation gives

$$x_n = \frac{r^n - 1}{r^n + 1}.$$

Since $|r| < 1$, then $\lim_{n \to \infty} r^n = 0$, and we have $\lim_{n \to \infty} x_n = -1$. Finally, if $|\alpha| > |\beta|$, then we use

$$\frac{\alpha^n - \beta^n}{\alpha^n + \beta^n} = -\frac{\beta^n - \alpha^n}{\beta^n + \alpha^n}$$

and the same argument given before will imply

$$\lim_{n \to \infty} x_n = -\lim_{n \to \infty} \frac{\beta^n - \alpha^n}{\beta^n + \alpha^n} = 1.$$

Solution 3.15

Let $\varepsilon > 0$. Since $\lim_{n \to \infty} x_n = l$, there exists $N_0 \geq 1$ such that for any $n \geq N_0$ we have

$$|x_n - l| < \frac{\varepsilon}{2}.$$

CHAPTER 3. SEQUENCES 53

On the other hand, we have

$$y_n - l = \frac{x_1 + x_2 + \cdots + x_n}{n} - l = \frac{(x_1 - l) + (x_2 - l) + \cdots + (x_n - l)}{n}$$

or

$$y_n - l = \frac{(x_1 - l) + (x_2 - l) + \cdots + (x_{N_0-1} - l)}{n} + \frac{(x_{N_0} - l) + \cdots + (x_n - l)}{n}$$

for any $n \geq N_0$. Since

$$\lim_{n \to \infty} \frac{(x_1 - l) + (x_2 - l) + \cdots + (x_{N_0-1} - l)}{n} = 0.$$

Then, there exists $N_1 \geq 1$ such that

$$\left| \frac{(x_1 - l) + (x_2 - l) + \cdots + (x_{N_0-1} - l)}{n} \right| < \frac{\varepsilon}{2}$$

for any $n \geq N_1$. Set $N \max\{N_0, N_1\}$, then for any $n \geq N$ we have

$$|y_n - l| \leq \left| \frac{(x_1 - l) + (x_2 - l) + \cdots + (x_{N_0-1} - l)}{n} \right| + \left| \frac{(x_{N_0} - l) + \cdots + (x_n - l)}{n} \right|$$

or

$$|y_n - l| \leq \left| \frac{(x_1 - l) + (x_2 - l) + \cdots + (x_{N_0-1} - l)}{n} \right| + \frac{|x_{N_0} - l| + \cdots + |x_n - l|}{n}$$

which implies

$$|y_n - l| < \frac{\varepsilon}{2} + \frac{n - N_0}{n} \frac{\varepsilon}{2} < \varepsilon.$$

This completes the proof of our statement. For the converse take $x_n = (-1)^n$. Then we have

$$y_n = \begin{cases} -\dfrac{1}{n} & \text{if } n \text{ is odd,} \\ 0 & \text{if } n \text{ is even.} \end{cases}$$

Obviously this will imply that $\lim_{n \to \infty} y_n = 0$ while $\{x_n\}$ is well known to be divergent. Finally, let $\{x_n\}$ be a sequence such that $\lim_{n \to \infty} x_{n+1} - x_n = l$. Set

$$y_n = \frac{(x_2 - x_1) + (x_3 - x_2) + \cdots + (x_{n+1} - x_n)}{n}.$$

Then from the first part we have $\lim_{n \to \infty} y_n = l$. But

$$y_n = \frac{x_{n+1} - x_1}{n}$$

which implies $x_{n+1} = n y_n + x_1$. Hence

$$\frac{x_n}{n} = \frac{n-1}{n} y_{n-1} + \frac{x_1}{n}.$$

Since

$$\lim_{n \to \infty} \frac{n-1}{n} = 1 , \ \lim_{n \to \infty} y_n = l , \ \text{and} \ \lim_{n \to \infty} \frac{x_1}{n} = 0$$

we get
$$\lim_{n \to \infty} \frac{x_n}{n} = l \ .$$

Solution 3.16

Assume first that $|l| < 1$. Let $\varepsilon = \dfrac{1 - |l|}{2}$. Then we have $\varepsilon > 0$. Since
$$\lim_{n \to \infty} \frac{x_{n+1}}{x_n} = l$$
we get
$$\lim_{n \to \infty} \left| \frac{x_{n+1}}{x_n} \right| = |l| \ .$$
Thus there exists $N_0 \geq 1$ such that for any $n \geq N_0$
$$\left| \frac{|x_{n+1}|}{|x_n|} - |l| \right| < \varepsilon$$
which implies
$$|l| - \varepsilon < \frac{|x_{n+1}|}{|x_n|} < |l| + \varepsilon$$
for any $n \geq N_0$. By definition of ε we get
$$\frac{|x_{n+1}|}{|x_n|} < \frac{|l| + 1}{2} < 1 \ .$$
In particular, we have for any $n \geq N_0$
$$|x_{n+1}| < \left(\frac{|l| + 1}{2} \right)^{n - N_0 + 1} |x_{N_0}|.$$
Since $\lim_{n \to \infty} \left(\dfrac{|l| + 1}{2} \right)^{n - N_0 + 1} = 0$, we get $\lim_{n \to \infty} |x_n| = 0$ which obviously implies $\lim_{n \to \infty} x_n = 0$. This completes the proof of the first part. Now assume $|l| > 1$. Since again
$$\lim_{n \to \infty} \left| \frac{x_{n+1}}{x_n} \right| = |l| \ ,$$
the same proof as above gives the existence of $N_0 \geq 1$ such that
$$\left(\frac{|l| + 1}{2} \right)^{n - N_0 + 1} |x_{N_0}| < |x_{n+1}|$$
for any $n \geq N_0$. And since $\lim_{n \to \infty} \left(\dfrac{|l| + 1}{2} \right)^{n - N_0 + 1} = \infty$, we get $\lim_{n \to \infty} |x_n| = \infty$. Hence the sequence $\{x_n\}$ is not bounded and therefore is divergent. Finally if we assume $|l| = 1$, then it is possible that $\{x_n\}$ may be convergent or divergent. For example, take $x_n = n^\alpha$, then we have $l = 1$. But the sequence only converges if $\alpha \leq 0$, otherwise it diverges. For the sequences
$$x_n = \frac{a^n}{n^k} \quad \text{and} \quad y_n = \frac{a^n}{n!} \ ,$$

CHAPTER 3. SEQUENCES

we have
$$\frac{x_{n+1}}{x_n} = \alpha \left(\frac{n}{n+1}\right)^k \text{ and } \frac{y_{n+1}}{y_n} = \alpha \frac{n!}{(n+1)!} = \alpha \frac{1}{n+1}.$$

Hence
$$\lim_{n\to\infty} \frac{x_{n+1}}{x_n} = \alpha \text{ and } \lim_{n\to\infty} \frac{y_{n+1}}{y_n} = 0.$$

In particular, we have
$$\begin{cases} \lim_{n\to\infty} x_n = 0 & \text{if } |\alpha| < 1, \\ \{x_n\} \text{ is divergent} & \text{if } |\alpha| > 1. \end{cases}$$

And if $|\alpha| = 1$, then the sequence in question is $\left\{\frac{1}{n^k}\right\}$ or $\left\{\frac{(-1)^n}{n^k}\right\}$ which is easy to conclude. For the sequence $\{y_n\}$ we have $\lim_{n\to\infty} y_n = 0$ regardless of the value of α.

Solution 3.17

Without loss of generality, we may assume $1 < x$. First note that
$$0 < \left(\sqrt[n]{x} - 1\right)^2 = \sqrt[n]{x^2} - 2\sqrt[n]{x} + 1,$$

which implies $2\sqrt[n]{x} - 1 < \sqrt[n]{x^2}$. Hence
$$\left(2\sqrt[n]{x} - 1\right)^n < \left(\sqrt[n]{x^2}\right)^n = x^2.$$

On the other hand, we have
$$\left(2\sqrt[n]{x} - 1\right)^n = x^2 \left(\frac{2\sqrt[n]{x} - 1}{\sqrt[n]{x^2}}\right)^n = x^2 \left(\frac{2}{\sqrt[n]{x}} - \frac{1}{\sqrt[n]{x^2}}\right)^n = x^2 \left(1 - \left(1 - \frac{1}{\sqrt[n]{x}}\right)^2\right)^n.$$

Since $(1-h)^n \geq 1 - nh$, for any $h \geq 0$ and $n \geq 1$ we get
$$\left(1 - \left(1 - \frac{1}{\sqrt[n]{x}}\right)^2\right)^n \geq 1 - n\left(1 - \frac{1}{\sqrt[n]{x}}\right)^2,$$

and
$$x = \left(\sqrt[n]{x} - 1 + 1\right)^n \geq 1 + n\left(\sqrt[n]{x} - 1\right) > n\left(\sqrt[n]{x} - 1\right),$$

which implies
$$\left(\sqrt[n]{x} - 1\right)^2 < \frac{x^2}{n^2}.$$

Hence
$$\left(2\sqrt[n]{x} - 1\right)^n \geq x^2 \left(1 - n\left(1 - \frac{1}{\sqrt[n]{x}}\right)^2\right) = x^2 \left(1 - n\frac{(\sqrt[n]{x} - 1)^2}{\sqrt[n]{x^2}}\right),$$

or
$$\left(2\sqrt[n]{x} - 1\right)^n > x^2 \left(1 - \frac{x^2}{n\sqrt[n]{x^2}}\right).$$

Putting all the inequalities together we get

$$x^2\left(1 - \frac{x^2}{n\sqrt[n]{x^2}}\right) < \left(2\sqrt[n]{x} - 1\right)^n < x^2 \; .$$

The Squeeze Theorem will then imply

$$\lim_{n\to\infty}\left(2\sqrt[n]{x} - 1\right)^n = x^2 \; ,$$

since

$$\lim_{n\to\infty} x^2\left(1 - \frac{x^2}{n\sqrt[n]{x^2}}\right) = x^2 \; .$$

Solution 3.18

In the previous problem we showed

$$x^2\left(1 - n\frac{\left(\sqrt[n]{x} - 1\right)^2}{\sqrt[n]{x^2}}\right) < \left(2\sqrt[n]{x} - 1\right)^n < x^2 \; ,$$

for any $x > 1$ and $n \geq 1$. Take $x = n$, we get

$$n^2\left(1 - n\frac{\left(\sqrt[n]{n} - 1\right)^2}{\sqrt[n]{n^2}}\right) \leq \left(2\sqrt[n]{n} - 1\right)^n \leq n^2 \; ,$$

which implies

$$1 - n\frac{\left(\sqrt[n]{n} - 1\right)^2}{\sqrt[n]{n^2}} \leq \frac{\left(2\sqrt[n]{n} - 1\right)^n}{n^2} \leq 1 \; .$$

In order to complete the proof of our statement we only need to show

$$\lim_{n\to\infty} n\frac{\left(\sqrt[n]{n} - 1\right)^2}{\sqrt[n]{n^2}} = 0 \; .$$

Note that for $x \in [0, 1]$ we have $0 \leq e^x - 1 \leq 3x$. Hence

$$0 \leq \sqrt[n]{n} - 1 = e^{\frac{\ln(n)}{n}} - 1 \leq 3\frac{\ln(n)}{n} \; ,$$

because $\ln(n) \leq n$ for $n \geq 1$. So

$$0 \leq n\left(\sqrt[n]{n} - 1\right)^2 \leq n 9\frac{\ln(n)^2}{n^2} = 9\frac{\ln^2(n)}{n} \; .$$

Since $\lim_{n\to\infty} \frac{\ln^2(n)}{n} = 0$, we conclude that

$$\lim_{n\to\infty} n\left(\sqrt[n]{n} - 1\right)^2 = 0 \; ,$$

CHAPTER 3. SEQUENCES

which yields
$$\lim_{n\to\infty} n\frac{\left(\sqrt[n]{n}-1\right)^2}{\sqrt[n]{n^2}} = 0 \ .$$

Solution 3.19

Let us first show by induction that $0 \leq x_n$ and $1 \leq x_n^2 \leq 2$. Obviously we have $0 \leq 1$ and $1 \leq 1^2 \leq 2$. Assume that $0 \leq x_n$ and $1 \leq x_n^2 \leq 2$. Then by the definition of x_{n+1} we obtain easily $0 \leq x_{n+1}$. On the other hand, we have

$$x_{n+1}^2 = \frac{1}{4}\left(x_n^2 + 4 + \frac{4}{x_n^2}\right) = \frac{1}{4}\left(x_n^2 + \frac{4}{x_n^2}\right) + 1 \ .$$

Since $(2-x_n^2)^2 = 4 - 4x_n^2 + x_n^4 \geq 0$ we get $\dfrac{x_n^4+4}{4x_n^2} \leq 1$ or $\dfrac{1}{4}\left(x_n^2 + \dfrac{4}{x_n^2}\right) \leq 1$. This will imply $x_{n+1}^2 \leq 1+1 = 2$. So the induction argument gives the desired conclusion that is $x_n \geq 0$ and $1 \leq x_n^2 \leq 2$, for any $n \geq 1$. On the other hand, algebraic manipulations give

$$x_{n+1} - x_n = \frac{1}{2}\left(x_n + \frac{2}{x_n}\right) - x_n = \frac{2 - x_n^2}{2x_n}$$

which implies $x_{n+1} - x_n \geq 0$ for any $n \geq 1$. Hence $\{x_n\}$ is an increasing bounded sequence. So it converges. Set $\lim_{n\to\infty} x_n = l$. Then we have $l \geq 0$ and $1 \leq l^2 \leq 2$. Since $\{x_{n+1}\}$ also converges to l, we get

$$l = \frac{1}{2}\left(l + \frac{2}{l}\right) = \frac{l^2+2}{2l} \ ,$$

or $2l^2 = l^2 + 2$, which gives $l^2 = 2$ or $l = \sqrt{2}$. Note that the sequence $\{x_n\}$ is formed of rational numbers and its limit is irrational. One may generalize this problem to the sequence

$$x_1 = 1 \text{ and } x_{n+1} = \frac{1}{2}\left(x_n + \frac{\alpha}{x_n}\right)$$

and show that $\{x_n\}$ converges to $\sqrt{\alpha}$ provided $\alpha \geq 0$.

Solution 3.20

Obviously the sequence $\{x_n\}$ is positive and since $x_{n+1} = \sqrt{x_n^2 + \frac{1}{2^n}} \geq \sqrt{x_n^2} = x_n$ in other words, the sequence $\{x_n\}$ is increasing. So in particular we have $x_n \geq x_1 = 1$ for any $n \geq 1$. Since

$$x_{n+1} - x_n = \sqrt{x_n^2 + \frac{1}{2^n}} - x_n = \frac{\frac{1}{2^n}}{\sqrt{x_n^2 + \frac{1}{2^n}} + x_n}$$

and

$$\sqrt{x_n^2 + \frac{1}{2^n}} + x_n \geq \sqrt{x_n^2} + x_n \geq \sqrt{1} + 1 = 2$$

we get
$$0 \leq x_{n+1} - x_n = \frac{\frac{1}{2^n}}{\sqrt{x_n^2 + \frac{1}{2^n}} + x_n} \leq \frac{1}{2^{n+1}}.$$

On the other hand, we have
$$x_{n+h} - x_n = (x_{n+h} - x_{n+h-1}) + (x_{n+h-1} - x_{n+h-2}) + \cdots + (x_{n+1} - x_n)$$
so
$$x_{n+h} - x_n \leq \frac{1}{2^{n+h}} + \frac{1}{2^{n+h-1}} + \cdots + \frac{1}{2^{n+1}} = \frac{1}{2^{n+1}}\left(\frac{1}{2^{h-1}} + \cdots + \frac{1}{2} + 1\right)$$
which implies
$$x_{n+h} - x_n \leq \frac{1}{2^{n+1}}\left(\frac{1 - \frac{1}{2^h}}{1 - \frac{1}{2}}\right) \leq \frac{1}{2^n}.$$

Since $\{\frac{1}{2^n}\}$ converges to 0, then for any $\varepsilon > 0$, there exists $N_0 \geq 1$ such that for any $n \geq N_0$, we have $\frac{1}{2^n} < \varepsilon$ which implies $x_{n+h} - x_n < \varepsilon$ for any $n \geq N_0$ and any $h \geq 1$. This obviously implies that $\{x_n\}$ is Cauchy. Therefore, $\{x_n\}$ is convergent. Note that if we are able to prove that $\{x_n\}$ is bounded, then we will get again the same conclusion without the complicated algebraic calculations.

Solution 3.21

1. One can easily show that $I_0 = \pi/2$ and $I_1 = 1$. For $n \geq 2$, we use the integration by parts technique to show
$$I_{n+2} = \int_0^{\pi/2} \cos^{n+1}(t)\cos(t)dt = \left[\cos^{n+1}(t)\sin(t)\right]_0^{\pi/2} + (n+1)\int_0^{\pi/2} \cos^n(t)\sin^2(t)dt,$$
which implies $I_{n+2} = (n+1)\bigl(I_n - I_{n+2}\bigr)$ or
$$I_{n+2} = \frac{n+1}{n+2} I_n.$$
Hence
$$I_{2n} = \frac{2n-1}{2n} \cdot \frac{2n-3}{2n-2} \cdots \frac{1}{2} I_0 = \frac{2n-1}{2n} \cdot \frac{2n-3}{2n-2} \cdots \frac{1}{2} \cdot \frac{\pi}{2} = \frac{(2n)!\pi}{2^{2n+1}(n!)^2},$$
and
$$I_{2n+1} = \frac{2n}{2n+1} \cdot \frac{2n-2}{2n-1} \cdots \frac{2}{3} I_1 = \frac{2n}{2n+1} \cdot \frac{2n-2}{2n-1} \cdots \frac{2}{3} = \frac{2^{2n}(n!)^2}{(2n+1)!}.$$

2. Note that since $0 \leq \cos^{n+1}(t) \leq \cos^n(t)$, for any $t \in [0, \pi/2]$, then $I_{n+1} \leq I_n$, i.e., $\{I_n\}$ is decreasing. In particular, we have $I_{n+2} \leq I_{n+1} \leq I_n$ and since $I_n > 0$ we get
$$1 \leq \frac{I_{n+1}}{I_{n+2}} \leq \frac{I_n}{I_{n+2}} = \frac{n+2}{n+1}.$$
Hence $\lim_{n \to \infty} \frac{I_{n+1}}{I_n} = 1$.

CHAPTER 3. SEQUENCES

3. Since
$$(n+2)I_{n+1}I_{n+2} = (n+1)I_n I_{n+1}$$
we conclude that $\{(n+1)I_n I_{n+1}\}$ is a constant sequence. Hence
$$(n+1)I_n I_{n+1} = I_0 I_1 = \frac{\pi}{2},$$
which implies $\lim_{n\to\infty} 2n I_n^2 = \lim_{n\to\infty} 2(n+1)I_n I_{n+1} = \pi$, or
$$\lim_{n\to\infty} I_n \sqrt{2n} = \sqrt{\pi}.$$

Solution 3.22

1. Note that $x_n > 0$ for $n \geq 1$. We have
$$\ln(x_{n+1}) - \ln(x_n) = \ln\left(\frac{x_{n+1}}{x_n}\right) = \ln\left(\frac{(n+1)!}{n!} \cdot \sqrt{\frac{n}{n+1}} \cdot e \cdot \frac{n^n}{(n+1)^{n+1}}\right)$$
which leads to
$$\ln(x_{n+1}) - \ln(x_n) = 1 - \left(n + \frac{1}{2}\right) \ln\left(1 + \frac{1}{n}\right).$$
Note that we have
$$\lim_{n\to\infty} n^2 \Big(\ln(x_{n+1}) - \ln(x_n) \Big) = \frac{1}{12}.$$
Indeed, using the Taylor approximation of $\ln(1+x)$ we get
$$\ln\left(1 + \frac{1}{n}\right) = \frac{1}{n} - \frac{1}{2n^2} + \frac{1}{6n^3} + \frac{\varepsilon_n}{n^3}$$
where $\{\varepsilon_n\}$ goes to 0 when $n \to \infty$. Hence
$$\ln(x_{n+1}) - \ln(x_n) = 1 - \left(n + \frac{1}{2}\right)\left(\frac{1}{n} - \frac{1}{2n^2} + \frac{1}{6n^3} + \frac{\varepsilon_n}{n^3}\right) = -\frac{1}{6n^2} + \frac{1}{4n^2} - \frac{\varepsilon_n}{n^2} - \frac{\varepsilon_n}{2n^3}$$
which implies
$$\lim_{n\to\infty} n^2 \Big(\ln(x_{n+1}) - \ln(x_n) \Big) = -\frac{1}{6} + \frac{1}{4} = \frac{1}{12}.$$
Since the series $\sum 1/n^2$ is convergent, the limit test will force $\sum \ln(x_{n+1}) - \ln(x_n)$ to be convergent. Hence $\ln(x_n)$ is convergent which in turn will force $\{x_n\}$ to be convergent. Set $l = \lim_{n\to\infty} x_n = e^L$, where $L = \lim_{n\to\infty} \ln(x_n)$. In particular, we have $l > 0$.

2. From the first part, we get
$$n! \approx l \left(\frac{n}{e}\right)^n \sqrt{n}, \text{ when } n \to \infty.$$

Using Wallis integrals (see Problem 3.21), $I_n = \int_0^{\pi/2} \cos^n(t)dt$, we know that $\lim_{n\to\infty} I_n \sqrt{2n} = \sqrt{\pi}$, or
$$I_n \approx \sqrt{\frac{\pi}{2n}}, \text{ when } n \to \infty.$$

Since $I_{2n} = \dfrac{(2n)!\pi}{2^{2n+1}(n!)^2}$, we get

$$\sqrt{\frac{\pi}{4n}} \approx \frac{(2n)!\pi}{2^{2n+1}(n!)^2}, \quad \text{when } n \to \infty,$$

which implies

$$\sqrt{\frac{\pi}{4n}} \approx \frac{l(2n)^{2n}e^{-2n}\sqrt{2n}}{2^{2n}(l n^n e^{-n}\sqrt{n})^2}\frac{\pi}{2}, \quad \text{when } n \to \infty.$$

Easy algebraic manipulations will lead to $l = \sqrt{2\pi}$.

3. Putting all the above results together we get

$$n! \approx \left(\frac{n}{e}\right)^n \sqrt{2\pi n}, \quad \text{when } n \to \infty.$$

Solution 3.23

- Notice that for any fixed n, $x_n = 2 + \frac{1}{2^n}$ if n is even and $x_n = \frac{1}{2^n}$ if n is odd. Thus $y_n = \sup\{x_n : k \geq n\} = 2 + \frac{1}{2^n}$ if n is even and $2 + \frac{1}{2^{n+1}}$ if n is odd. Hence

$$\limsup\{x_n\} = \inf\{y_n : n \in \mathbb{N}\} = 2.$$

A similar calculation yields $\liminf\{x_n\} = 0$.

- Because $\{x_n\}$ is not bounded above, the limit superior does not exist. For the limit inferior, consider $z_n = \inf\{x_k : k \geq n\}$. Clearly, $z_n = x_n = 2^n$, since $\{x_n\}$ is monotone increasing and z_n diverges to ∞. Thus supremum over $\{z_n : n \in \mathbb{N}\}$ does not exist, therefore the limit inferior does not exist. Note that even though the sequence $\{x_n\}$ is bounded below, limit inferior does not exist.

Solution 3.24

Since

$$\liminf_{n\to\infty} -x_n = -\limsup_{n\to\infty} x_n,$$

we will only prove the existence of a subsequence which converges to $\liminf\limits_{n\to\infty} x_n$. It is clear that $\liminf\limits_{n\to\infty} x_n = l \in \mathbb{R}$ since $\{x_n\}$ is bounded below. For any $\varepsilon > 0$ there exists $N \in \mathbb{N}$, such that for any $n \geq N$ we have

$$l - \varepsilon < \inf\{x_k; k \geq n\} \leq l.$$

Set $\varepsilon = 1$, then there exists $N_1 \in \mathbb{N}$ such that for any $n \geq N_1$ we have

$$l - 1 < \inf\{x_k; k \geq n\} \leq l.$$

By induction one will construct an increasing sequence of integers $\{N_i\} \in \mathbb{N}$ such that for any $n \geq N_i$ we have

$$l - \frac{1}{i} < \inf\{x_k; k \geq n\} \leq l.$$

CHAPTER 3. SEQUENCES 61

In particular, we have $l - 1/k < x_{N_k} \leq l$, which implies $\{x_{N_k}\} \to l$.

Solution 3.25

Note that for any sequence $\{x_n\}$ we have $\liminf\limits_{n \to \infty} x_n \leq \limsup\limits_{n \to \infty} x_n$. Since $\liminf\limits_{n \to \infty} -x_n = -\limsup\limits_{n \to \infty} x_n$, we will only show that $\liminf\limits_{n \to \infty} x_n \leq \liminf\limits_{n_k \to \infty} x_{n_k}$. By definition we have

$$\inf\{x_k; k \geq n\} \leq \inf\{x_{n_k}; n_k \geq n\}, \ n \in \mathbb{N}.$$

Hence
$$\inf\{x_k; k \geq n'\} \leq \sup_{n \in \mathbb{N}} \Big(\inf\{x_{n_k}; n_k \geq n\} \Big), \ n' \in \mathbb{N},$$

or
$$\sup_{n' \in \mathbb{N}} \Big(\inf\{x_k; k \geq n'\} \Big) \leq \sup_{n \in \mathbb{N}} \Big(\inf\{x_{n_k}; n_k \geq n\} \Big)$$

which implies $\liminf\limits_{n \to \infty} x_n \leq \liminf\limits_{n_k \to \infty} x_{n_k}$. Moreover if we assume that $\{x_{n_k}\}$ is convergent, then we have
$$\liminf_{n_k \to \infty} x_{n_k} = \limsup_{n_k \to \infty} x_{n_k} = \lim_{n_k \to \infty} x_{n_k},$$

which implies $\liminf\limits_{n \to \infty} x_n \leq \lim\limits_{n_k \to \infty} x_{n_k} \leq \limsup\limits_{n \to \infty} x_n$. The converse is not true. Indeed, consider the sequence $\{(-1)^n\}$. Then we have $\liminf\limits_{n \to \infty}(-1)^n = -1$ and $\limsup\limits_{n \to \infty}(-1)^n = 1$. On other hand there does not exist a subsequence which converges to 0.

Solution 3.26

For any $N \in \mathbb{N}$, we have
$$x_n + y_n \leq \sup\{x_k; \ k \geq N\} + \sup\{y_k; \ k \geq N\}, \ n \geq N$$

which implies $\sup\{x_n + y_n; \ n \geq N\} \leq \sup\{x_k; \ k \geq N\} + \sup\{y_k; \ k \geq N\}$. Hence

$$\inf_{N \in \mathbb{N}} \Big(\sup\{x_n + y_n; \ n \geq N\} \Big) \leq \inf_{N \in \mathbb{N}} \Big(\sup\{x_n; \ n \geq N\} \Big) + \inf_{N \in \mathbb{N}} \Big(\sup\{y_n; \ n \geq N\} \Big),$$

or $\limsup\limits_{n \to \infty}(x_n + y_n) \leq \limsup\limits_{n \to \infty} x_n + \limsup\limits_{n \to \infty} y_n$. The equality does not hold in general. Indeed, we have $\limsup\limits_{n \to \infty}(-1)^n = 1$, and $\limsup\limits_{n \to \infty}(-1)^{n+1} = 1$, but $\limsup\limits_{n \to \infty}(-1)^n + (-1)^{n+1} = 0$.

Solution 3.27

Assume first that $\liminf\limits_{n \to \infty} \dfrac{x_{n+1}}{x_n} = l \in \mathbb{R}$. So for any $\varepsilon > 0$, there exists $N \in \mathbb{N}$ such that for any $n \geq N$, we have $l - \varepsilon \leq \inf\limits_{n \geq N} \dfrac{x_{n+1}}{x_n}$, which implies $(l - \varepsilon)x_n \leq x_{n+1}$ for any $n \geq N$. This clearly implies $(l - \varepsilon)^{n-N} x_N \leq x_n$, for any $n \geq N$. Hence

$$(l - \varepsilon)^{(n-N)/n} x_N^{1/n} \leq x_n^{1/n}.$$

Since $(l - \varepsilon)^{(n-N)/n} x_N^{1/n} \to (l - \varepsilon)$ when $n \to \infty$, we get

$$l - \varepsilon \leq \liminf_{n \to \infty} x_n^{1/n}.$$

Since ε was arbitrarily positive, we get

$$\liminf_{n\to\infty} \frac{x_{n+1}}{x_n} \leq \liminf_{n\to\infty} \sqrt[n]{x_n}.$$

A similar proof will lead to

$$\limsup_{n\to\infty} \sqrt[n]{x_n} \leq \limsup_{n\to\infty} \frac{x_{n+1}}{x_n}.$$

If $\{x_{n+1}/x_n\}$ is convergent, then we have

$$\liminf_{n\to\infty} \frac{x_{n+1}}{x_n} = \limsup_{n\to\infty} \frac{x_{n+1}}{x_n},$$

which obviously implies

$$\liminf_{n\to\infty} \sqrt[n]{x_n} = \limsup_{n\to\infty} \sqrt[n]{x_n} = \lim_{n\to\infty} \frac{x_{n+1}}{x_n} = \lim_{n\to\infty} \sqrt[n]{x_n}.$$

The converse is not true. Indeed, take $x_n = 2 + (-1)^n$, $n \in \mathbb{N}$. It is easy to check that $\sqrt[n]{x_n} \to 1$ when $n \to \infty$. But

$$\liminf_{n\to\infty} \frac{x_{n+1}}{x_n} = \frac{1}{3}, \text{ and } \limsup_{n\to\infty} \frac{x_{n+1}}{x_n} = 3.$$

Chapter 4
Limits of Functions

When a variable quantity converges towards a fixed limit, it is often useful to indicate this limit by a specific notation, which we shall do by setting the abbreviation

lim

in front of the variable in question.

Augustin Louis Cauchy (1789–1857)

- Let $f : D \to \mathbb{R}$ and let c be an accumulation point of D. We say that a real number L is a *limit of f at c*, and write
$$\lim_{x \to c} f(x) = L,$$
if for each $\varepsilon > 0$ there exists a $\delta > 0$ such that $|f(x) - L| < \varepsilon$ for all points $x \in D$ for which $0 < |x - c| < \delta$.

- *Monotone function*: A function $f : A \to \mathbb{R}$ is *increasing* on A if $f(x) \leq f(y)$ whenever $x < y$ and *decreasing* if $f(x) \geq f(y)$ whenever $x < y$ in A. A *monotone* function is one that is either increasing or decreasing.

- *Bounded function*: Let $f : A \to \mathbb{R}$ be a function and $B \subseteq A$. We say f is bounded on B if $f(B)$ is bounded, where $f(B)$ is the range of f over B, i.e., $f(B) = \{f(x) : x \in B\}$.

- *One-sided limits*: Suppose the domain of f is an interval (a,b), then the *right-hand limit* of $f(x)$ at a written by $\lim_{x \to a^+} f(x)$ and $\lim_{x \to a^+} f(x) = L$ if and only if for every $\varepsilon > 0$ there exists a $\delta > 0$ such that $|f(x) - L| < \varepsilon$ whenever $x \in (a,b)$ and $a < x < a + \delta$. Similarly, the left-hand limit of f at b is given by $\lim_{x \to b^-} f(x) = L$ if and only if for every $\varepsilon > 0$ there exists a $\delta > 0$ such that $|f(x) - L| < \varepsilon$ whenever $x \in (a,b)$ and $b - \delta < x < b$. Of course, $\lim_{x \to c} f(x) = L$ if and only if both one-sided limits exist and are equal to L.

- If f is an increasing function on the interval (a,b), then one-sided limits of f exist at each point $c \in (a,b)$, and
$$\lim_{x \to c^-} f(x) = L \leq f(c) \leq \lim_{x \to c^+} f(x) = M.$$

  For decreasing functions, the above inequalities are reversed.

- Let $f : (b, \infty) \to \mathbb{R}$, where $b \in \mathbb{R}$. We say that $L \in \mathbb{R}$ is the limit of f as $x \to \infty$, and we write
$$\lim_{x \to \infty} f(x) = L,$$
provided that for each $\varepsilon > 0$ there exists a real number $N > b$ such that $x > N$ implies that $|f(x) - L| < \varepsilon$.

- Let $f : (b, \infty) \to \mathbb{R}$. We say that f *tends to* ∞ as $x \to \infty$, and we write
$$\lim_{x \to \infty} f(x) = \infty,$$
provided that given any $M \in \mathbb{R}$ there exists a $N > b$ such that $x > N$ implies that $f(x) > M$.

Problem 4.1 Let $f : D \to \mathbb{R}$. Let $x_0 \in D$ such that $\lim_{x \to x_0} f(x)$ exists. Show that $\lim_{x \to x_0} |f(x)|$ exists and the following identity holds:
$$\lim_{x \to x_0} |f(x)| = \left| \lim_{x \to x_0} f(x) \right|.$$

Problem 4.2 Consider the function
$$f(x) = \begin{cases} 0 & \text{if } x \in \mathbb{R} \text{ and } x \text{ is rational,} \\ 1 & \text{if } x \in \mathbb{R} \text{ and } x \text{ is irrational.} \end{cases}$$

Show that $\lim_{x \to a} f(x)$ does not exist for any $a \in \mathbb{R}$.

CHAPTER 4. LIMITS OF FUNCTIONS

Problem 4.3 (Dirichlet Function) Consider the function $f : [0, 1] \to \mathbb{R}$ defined by

$$f(x) = \begin{cases} \dfrac{1}{q} & \text{if } x \text{ is rational and } x = \dfrac{p}{q} \text{ in the reducible form,} \\ 0 & \text{if } x \in \mathbb{R} \text{ and } x \text{ is irrational.} \end{cases}$$

Show that $\lim_{x \to a} f(x) = 0$ for any $a \in (0, 1)$.

Problem 4.4 Consider the function

$$f(x) = \begin{cases} x - 1 & \text{if } x \in \mathbb{R} \text{ is rational,} \\ 5 - x & \text{if } x \in \mathbb{R} \text{ is irrational.} \end{cases}$$

Show that $\lim_{x \to 3} f(x)$ exists but $\lim_{x \to a} f(x)$ does not exist for any $a \neq 3$.

Problem 4.5 Let $f(x)$ be a periodic function. Show that if $\lim_{x \to \infty} f(x)$ exists, then $f(x)$ is a constant function. Deduce from this that $\lim_{x \to \infty} \sin x$ does not exist.

Problem 4.6 Evaluate the following limits:

$$\lim_{x \to 4} \frac{4 - x}{2 - \sqrt{x}} \quad \text{and} \quad \lim_{x \to 0} x \sin\left(\frac{1}{x}\right).$$

Problem 4.7 Evaluate the limit

$$\lim_{x \to 0} \frac{\sin(x)}{\sqrt{1 - \cos(x)}}.$$

Problem 4.8 Let f be real-valued functions defined on I. Let $x_0 \in I$. Show that $\lim_{x \to x_0} f(x)$ exists if and only if for any sequence $\{x_n\}$ in I which converges to x_0, the sequence $\{f(x_n)\}$ is convergent.

Problem 4.9 (Squeeze Theorem) Let f, g, and h be three real-valued functions defined on $I \subset \mathbb{R}$. Assume that for all $x \in I$, we have
$$f(x) \leq g(x) \leq h(x)$$
and that for $x_0 \in I$ we have $\lim_{x \to x_0} f(x) = \lim_{x \to x_0} h(x) = l$. Prove that
$$\lim_{x \to x_0} g(x) = l \, .$$

Problem 4.10 Let $f, g : D \to \mathbb{R}$. Let $x_0 \in D$ such that $\lim_{x \to x_0} f(x)$ and $\lim_{x \to x_0} g(x)$ exist. Discuss the existence of the limits
$$\lim_{x \to x_0} \min\{f(x), g(x)\} \, , \text{ and } \lim_{x \to x_0} \max\{f(x), g(x)\} \, .$$

Problem 4.11 Let $f : [0,1] \to \mathbb{R}$ be a monotone function. Let $x_0 \in (0,1)$ and assume that $\lim_{x \to x_0} f(x)$ exists. Show that
$$\lim_{x \to x_0} f(x) = f(x_0) \, .$$

Problem 4.12 Let $f : [a,b] \to \mathbb{R}$ be monotone. Prove that

(i) $\lim_{x \to x_0+} f(x)$ and $\lim_{x \to x_0-} f(x)$ exist for any $x_0 \in (a,b)$;

(ii) $\lim_{x \to a+} f(x)$ and $\lim_{x \to b-} f(x)$ exist.

Problem 4.13 Let $f : [a,b] \to \mathbb{R}$ be monotone. Prove that the set
$$D = \{x : \ x \in [a,b] \text{ and } f \text{ does not have a limit at } x\}$$
is countable. What happens at the points $\{x\}$ that are not in D?

Problem 4.14 Let $f : \mathbb{R} \to \mathbb{R}$ such that
$$f(x+y) = f(x) + f(y)$$
for any $x, y \in \mathbb{R}$. Assume that $\lim_{x \to 0} f(x) = f(0)$. Find $f(x)$.

Problem 4.15 We say f is asymptotic to g and write $f \sim g$ if $\frac{f(x)}{g(x)} \to 1$ as $x \to \infty$. We write $f = O(g)$ (pronounced "Big Oh") if $g(x) > 0$ for sufficiently large $x \in \mathbb{R}$ and $\frac{f(x)}{g(x)}$ is bounded for sufficiently large x. We write $f = o(g)$ (pronounced "Little Oh") if $\frac{f(x)}{g(x)} \to 0$ as $x \to \infty$. Prove the following:

1. $x^2 + x \sim x^2$
2. $x^2 + x = O(x^2)$
3. $e^{\sqrt{\log x}} = o(x)$
4. $x^{\log x} = o(e^x)$

Problem 4.16 Prove that if $f = o(g)$ and if $g(x) \to \infty$ as $x \to \infty$, then $e^f = o(e^g)$ as $x \to \infty$.

Solutions

Solution 4.1

Set $\lim_{x \to x_0} f(x) = l$. Let us prove that $\lim_{x \to x_0} |f(x)| = |l|$. Let $\varepsilon > 0$. Then there exists $\delta > 0$ such that for any $x \in D$, $|f(x) - l| < \varepsilon$ provided $|x - x_0| < \delta$. Using the inequality

$$\Big||a| - |b|\Big| \leq |a - b|$$

for any real numbers a and b. Then for any $x \in D$ such that $|x - x_0| < \delta$, we have

$$\Big||f(x)| - |l|\Big| \leq |f(x) - l| < \varepsilon \;.$$

This completes the proof of our statement.

Solution 4.2

Let $a \in \mathbb{R}$. Assume that $\lim_{x \to a} f(x) = l$ exists. Then for $\varepsilon = \dfrac{1}{3}$, there exists $\delta > 0$ such that for any $x \in (a - \delta, a + \delta)$ we have $|f(x) - l| < \varepsilon$. We know that any nonempty open interval contains a rational and an irrational number. Let $x \in (a - \delta, a + \delta)$ be a rational number. Then we have $|f(x) - l| = |0 - l| < \varepsilon$. And if $x \in (a - \delta, a + \delta)$ is an irrational number, then we have $|f(x) - l| = |1 - l| < \varepsilon$. Hence

$$1 = |0 - 1| \leq |0 - l| + |1 - l| < 2\varepsilon = \frac{2}{3} \;.$$

This contradiction implies that $\lim_{x \to a} f(x)$ does not exist.

Solution 4.3

Let $a \in (0, 1)$. Fix $\varepsilon > 0$. Consider the set

$$A_\varepsilon = \left\{ \frac{p}{q} \in \mathbb{Q} \cap [0, 1]; \frac{1}{q} \geq \varepsilon \right\} \;.$$

The set A_ε is finite. Set

$$\delta = \min\left\{ \left|a - \frac{p}{q}\right|; \frac{p}{q} \in A_\varepsilon \text{ and } \frac{p}{q} \neq a \right\} \;.$$

Obviously we have $\delta > 0$. Let $x \in [0, 1]$ such that $|x - a| < \delta$. If x is irrational, then $|f(x) - 0| = 0 < \varepsilon$. Now assume that x is rational and $x = \dfrac{p}{q}$ is its reduced form, i.e., p and q are coprime. Since $|x - a| < \delta$, then $\dfrac{p}{q} \notin A_\varepsilon$. Hence $\dfrac{1}{q} < \varepsilon$ which implies

$$|f(x) - 0| = \left|\frac{1}{q} - 0\right| = \frac{1}{q} < \varepsilon \;.$$

Therefore, for any $x \in [0, 1]$ such that $|x - a| < \delta$, we have $|f(x) - 0| = 0 < \varepsilon$. This completes the proof of our problem.

Solution 4.4

Note that
$$\lim_{x \to 3} x - 1 = 2 \text{ and } \lim_{x \to 3} 5 - x = 2 .$$
So for any $\varepsilon > 0$, there exist $\delta_1 > 0$ and $\delta_2 > 0$ such that
$$\begin{cases} |x - 1 - 2| < \varepsilon & \text{if } x \in \mathbb{R} \text{ and } |x - 3| < \delta_1, \\ |5 - x - 2| < \varepsilon & \text{if } x \in \mathbb{R} \text{ and } |x - 3| < \delta_2. \end{cases}$$
Set $\delta = \min\{\delta_1, \delta_2\}$. Then we have $\delta > 0$. Let $x \in \mathbb{R}$ such that $|x - 3| < \delta$. If x is rational, then because $|x - 3| < \delta \leq \delta_1$, we get
$$|f(x) - 2| = |x - 1 - 2| < \varepsilon ,$$
and if x is irrational, then because $|x - 3| < \delta \leq \delta_2$, we get
$$|f(x) - 2| = |5 - x - 2| < \varepsilon .$$
So for any $x \in \mathbb{R}$ such that $|x - 3| < \delta$, we have $|f(x) - 2| < \varepsilon$. This clearly implies $\lim_{x \to 3} f(x) = 2$.
Finally let $a \in \mathbb{R}$ with $a \neq 3$. Assume that $\lim_{x \to a} f(x) = l$ exists. Let $\varepsilon = \dfrac{|3 - a|}{3}$. We have $\varepsilon > 0$. Then there exists $\delta > 0$ such that for any $x \in \mathbb{R}$ such that $|x - a| < \delta$, we have $|f(x) - l| < \varepsilon$. Set $\delta^* = \min\{\delta, \varepsilon\}$. The open interval $(a - \delta^*, a + \delta^*)$ contains rational and irrational numbers. Let $x \in (a - \delta^*, a + \delta^*)$ be rational. Then we have
$$|f(x) - l| = |x - 1 - l| < \varepsilon .$$
But
$$|a - 1 - l| \leq |a - x| + |x - 1 - l| < \delta^* + \varepsilon < 2\varepsilon .$$
On the other hand, if $x \in (a - \delta^*, a + \delta^*)$ is irrational, then we have
$$|f(x) - l| = |5 - x - l| < \varepsilon .$$
But
$$|5 - a - l| \leq |x - a| + |5 - x - l| < \delta^* + \varepsilon < 2\varepsilon .$$
Hence
$$|6 - 2a| = |5 - a - l - (a - 1 - l)| \leq |5 - a - l| + |a - 1 - l| < 4\varepsilon = \frac{4|3 - a|}{3},$$
or $|3 - a| < \dfrac{2|3 - a|}{3}$. This contradiction implies that $\lim_{x \to a} f(x) = l$ does not exist.

Solution 4.5

Assume that $\lim_{x \to \infty} f(x) = l$ exists. Let us show that $f(x) = l$ for any $x \in \mathbb{R}$. Let $T > 0$ be a period of $f(x)$. Let $a \in \mathbb{R}$ and assume that $f(a) \neq l$. Take $\varepsilon = \dfrac{|f(a) - l|}{2}$. Since $\lim_{x \to \infty} f(x) = l$, then there exists $M > 0$ such that for any $x > M$, we have $|f(x) - l| < \varepsilon$. Since $\mathbb{R}$ is Archimedean, there exists $n \in \mathbb{N}$ such that $n > \dfrac{M-a}{T}$. Then we have $a + nT > M$. Hence $|f(a+nT) - l| < \varepsilon$. Since $f(a+nT) = f(a)$, we get

$$|f(a) - l| < \varepsilon = \frac{|f(a) - l|}{2}$$

which is a contradiction, therefore $f(a) = l$. Since a was arbitrary, we obtain the conclusion of our claim. Since $\sin(x)$ is periodic and is not a constant function, this implies that it does not have a limit as x tends to ∞.

$\boxed{\text{Solution 4.6}}$

Let us first discuss $\lim\limits_{x \to 4} \dfrac{4-x}{2-\sqrt{x}}$. The following algebraic identity will be helpful:

$$\frac{a-b}{\sqrt{a}-\sqrt{b}} = \frac{(a-b)(\sqrt{a}+\sqrt{b})}{(\sqrt{a}-\sqrt{b})(\sqrt{a}+\sqrt{b})} = \frac{(a-b)(\sqrt{a}+\sqrt{b})}{a-b} = \sqrt{a} + \sqrt{b},$$

whenever $a \neq b$. Note that if x is close to 4, then we may assume that $x > 0$ and $x \neq 4$. Hence we have

$$\frac{4-x}{2-\sqrt{x}} = 2 + \sqrt{x}.$$

From this we suspect that $\lim\limits_{x \to 4} \dfrac{4-x}{2-\sqrt{x}} = 4$. Indeed, let $\varepsilon > 0$. Set $\delta = \min(2\varepsilon, 1)$. Let $x \in (4-\delta, 4+\delta)$ and not equal 4. Then $x \in (3,5)$ and is therefore positive. Since

$$\left| \frac{4-x}{2-\sqrt{x}} - 4 \right| = |2 + \sqrt{x} - 4| = |\sqrt{x} - 2| = \left| \frac{x-4}{\sqrt{x}+2} \right| \leq \frac{|x-4|}{2},$$

we get

$$\left| \frac{4-x}{2-\sqrt{x}} - 4 \right| < \frac{\delta}{2} \leq \frac{2\varepsilon}{2},$$

or

$$\left| \frac{4-x}{2-\sqrt{x}} - 4 \right| < \varepsilon.$$

This completes the proof of our claim. Next we discuss $\lim\limits_{x \to 0} x \sin\left(\dfrac{1}{x}\right)$. The first common mistake is to write

$$\lim_{x \to 0} x \sin\left(\frac{1}{x}\right) = \lim_{x \to 0} x \left[\lim_{x \to 0} \sin\left(\frac{1}{x}\right) \right] = 0$$

because the above identity holds if both limits exist. But we previously showed that $\sin(x)$ does not have a limit when $x \to \infty$ which implies the limit

$$\lim_{x \to 0+} \sin\left(\frac{1}{x}\right) = \lim_{t \to \infty} \sin(t)$$

does not exist, where $t = \dfrac{1}{x}$. In order to solve this problem note that $\sin(x)$ is always bounded and $\lim\limits_{x \to 0} x = 0$. So we are multiplying a bounded function with a function getting very small. The end

CHAPTER 4. LIMITS OF FUNCTIONS 71

result should be small. Hence we guess that $\lim_{x \to 0} x \sin\left(\frac{1}{x}\right) = 0$. Let us prove it. Let $\varepsilon > 0$. Take $\delta = \varepsilon$ and let $x \in \mathbb{R}$ such that $|x - 0| < \delta$ or $|x| < \delta$. Then we have

$$\left| x \sin\left(\frac{1}{x}\right) - 0 \right| = \left| x \sin\left(\frac{1}{x}\right) \right| \leq |x| < \delta = \varepsilon .$$

This completes the proof of our claim.

Solution 4.7

The square root in the denominator is the part complicating this limit. In order to get rid of this problem, we will use the following trigonometric identities:

$$\cos(\theta) = 1 - 2\sin^2\left(\frac{\theta}{2}\right) \quad \text{and} \quad \sin(\theta) = 2\cos\left(\frac{\theta}{2}\right)\sin\left(\frac{\theta}{2}\right) .$$

So

$$\frac{\sin(x)}{\sqrt{1 - \cos(x)}} = \frac{2\cos\left(\frac{x}{2}\right)\sin\left(\frac{x}{2}\right)}{\sqrt{2\sin^2\left(\frac{x}{2}\right)}} = \frac{2\cos\left(\frac{x}{2}\right)\sin\left(\frac{x}{2}\right)}{\sqrt{2}\left|\sin\left(\frac{x}{2}\right)\right|} .$$

First note that

$$\lim_{x \to 0} \frac{2\cos\left(\frac{x}{2}\right)}{\sqrt{2}} = \sqrt{2} .$$

So let us focus on the limit $\lim_{x \to 0} \frac{\sin\left(\frac{x}{2}\right)}{\left|\sin\left(\frac{x}{2}\right)\right|}$. In order to get rid of the absolute value, we will consider the limits to the right of 0 and to the left of 0. From the properties of the sin function we know that for any $x \in (0, \pi)$, then $\sin\left(\frac{x}{2}\right) > 0$ and if $x \in (-\pi, 0)$, then $\sin\left(\frac{x}{2}\right) < 0$. Hence

$$\lim_{x \to 0+} \frac{\sin\left(\frac{x}{2}\right)}{\left|\sin\left(\frac{x}{2}\right)\right|} = \lim_{x \to 0+} \frac{\sin\left(\frac{x}{2}\right)}{\sin\left(\frac{x}{2}\right)} = 1 ,$$

and

$$\lim_{x \to 0-} \frac{\sin\left(\frac{x}{2}\right)}{\left|\sin\left(\frac{x}{2}\right)\right|} = \lim_{x \to 0+} \frac{\sin\left(\frac{x}{2}\right)}{-\sin\left(\frac{x}{2}\right)} = -1 .$$

Putting all the previous information together we get

$$\lim_{x \to 0+} \frac{\sin(x)}{\sqrt{1 - \cos(x)}} = \sqrt{2} \quad \text{and} \quad \lim_{x \to 0-} \frac{\sin(x)}{\sqrt{1 - \cos(x)}} = -\sqrt{2} .$$

This obviously implies that the limit $\lim_{x \to 0} \frac{\sin(x)}{\sqrt{1 - \cos(x)}}$ does not exist.

Solution 4.8

Assume first that $\lim_{x \to x_0} f(x) = p$ exists. Let $\{x_n\}$ in I be a sequence which converges to x_0. Let us show that $\{f(x_n)\}$ converges to p. Indeed, let $\varepsilon > 0$. Then there exists $\delta > 0$ such that $|f(x) - p| < \varepsilon$ provided $|x - x_0| < \delta$ and $x \in I$. Since $\{x_n\}$ converges to x_0, then there exists $N \geq 1$ such that for any $n \geq N$ we have $|x_n - x_0| < \delta$. Putting everything together we get $|f(x_n) - p| < \varepsilon$ for any $n \geq N$. This proves our claim that $\{f(x_n)\}$ converges to p. Conversely, assume that for any sequence $\{x_n\}$ in I which converges to x_0, the sequence $\{f(x_n)\}$ is convergent. Let us show that $\lim_{x \to x_0} f(x)$ exists. First let us prove that there exists $p \in \mathbb{R}$ such that for any sequence $\{x_n\}$ in I which converges to x_0, the sequence $\{f(x_n)\}$ converges to p. Indeed, let $\{x_n\}$ and $\{y_n\}$ be two sequences in I which converges to x_0. Define the new sequence $\{z_n\}$ by

$$z_{2n} = x_n \text{ and } z_{2n+1} = y_n,$$

for all $n \in \mathbb{N}$. It is easy to check that $\{z_n\}$ is in I and converges to x_0. Hence our assumption implies that $\{f(z_n)\}$ is convergent. Since both $\{f(x_n)\}$ and $\{f(y_n)\}$ are subsequences of $\{f(z_n)\}$, they are both convergent and they have the same limit. This proves that the limit of $\{f(x_n)\}$ is independent of the sequence $\{x_n\}$. Let us call this limit p. Finally we need to show that $\lim_{x \to x_0} f(x) = p$. Assume not. Then there exists $\varepsilon_0 > 0$ such that for any $\delta > 0$ there exists $x_\delta \in I$ such that $|x_\delta - x_0| < \delta$ and $|f(x_\delta) - p| \geq \varepsilon_0$. If we let $\delta = 1/n$ for $n \geq 1$, we will generate a sequence $\{x_n\}$ in I such that

$$|x_n - x_0| < \frac{1}{n} \text{ and } |f(x_n) - p| \geq \varepsilon_0.$$

Hence $\{x_n\}$ converges to x_0 and $\{f(x_n)\}$ does not converge to p. This contradiction establishes our claim, i.e., $\lim_{x \to x_0} f(x) = p$.

Solution 4.9

Let $\varepsilon > 0$. Since $\lim_{x \to x_0} f(x) = \lim_{x \to x_0} h(x) = l$, there exist $\delta_1 > 0$ and $\delta_2 > 0$ such that

$$\begin{cases} |f(x) - l| < \varepsilon & \text{if } x \in I \text{ and } |x - x_0| < \delta_1, \\ |h(x) - l| < \varepsilon & \text{if } x \in I \text{ and } |x - x_0| < \delta_2. \end{cases}$$

Set $\delta = \min\{\delta_1, \delta_2\}$. Hence $\delta > 0$. Let $x \in I$ such that $|x - x_0| < \delta$. Then we have $|x - x_0| < \delta_1$ and $|x - x_0| < \delta_2$. Hence $|f(x) - l| < \varepsilon$ and $|h(x) - l| < \varepsilon$. In particular, we have

$$\begin{cases} l - \varepsilon < f(x) < l + \varepsilon, \\ l - \varepsilon < h(x) < l + \varepsilon. \end{cases}$$

Since $f(x) \leq g(x) \leq h(x)$ for any $x \in I$, we get

$$l - \varepsilon < g(x) < l + \varepsilon,$$

or $|g(x) - l| < \varepsilon$. This completes the proof of our statement.

CHAPTER 4. LIMITS OF FUNCTIONS

Solution 4.10

We will make use of the identities
$$\max\{a,b\} = \frac{a+b+|a-b|}{2} \quad \text{and} \quad \min\{a,b\} = \frac{a+b-|a-b|}{2},$$
for any $a, b \in \mathbb{R}$. Now set
$$\lim_{x \to x_0} f(x) = l \quad \text{and} \quad \lim_{x \to x_0} g(x) = L.$$
Properties on the limits of functions will imply $\lim_{x \to x_0} f(x) - g(x) = l - L$. In Problem 5.1, we showed that $\lim_{x \to x_0} \left| f(x) - g(x) \right| = |l - L|$. Hence
$$\lim_{x \to x_0} \frac{f(x) + g(x) + |f(x) - g(x)|}{2} = \frac{l + L + |l - L|}{2},$$
or
$$\lim_{x \to x_0} \max\{f(x), g(x)\} = \max\{l, L\}.$$
A similar proof will give
$$\lim_{x \to x_0} \min\{f(x), g(x)\} = \min\{l, L\}.$$
It is obvious that one may generalize these conclusions to a finite number of functions. But the infinite case is not true. Indeed, take
$$f_n(x) = \begin{cases} x^n & \text{if } x \in [0,1], \\ 1 & \text{if } x \geq 1. \end{cases}$$
Then it is quite easy to check that $\lim_{x \to 1} f_n(x) = 1$. But if we set $f(x) = \inf_{n \geq 1} f_n(x)$, then we have
$$f(x) = \begin{cases} 0 & \text{if } x \in [0,1], \\ 1 & \text{if } x \geq 1. \end{cases}$$
Hence
$$\lim_{x \to 1-} f(x) = 0 \quad \text{and} \quad \lim_{x \to 1+} f(x) = 1.$$

Solution 4.11

Without loss of generality, we may assume that $f(x)$ is increasing. Set $\lim_{x \to x_0} f(x) = l$. Let us prove that $l = f(x_0)$. Let $\varepsilon > 0$. Then there exists $\delta > 0$ such that for any $x \in [0,1]$ such that $|x_0 - x| < \delta$, we get $|f(x) - l| < \varepsilon$. In particular we have
$$\begin{cases} l - \varepsilon < f(x) < l + \varepsilon & \text{if } x \in [0,1] \text{ and } x_0 < x < x_0 + \delta, \\ l - \varepsilon < f(x) < l + \varepsilon & \text{if } x \in [0,1] \text{ and } x_0 - \delta < x < x_0. \end{cases}$$
Since $f(x)$ is increasing, for any $x \in [0,1]$ and $x_0 < x < x_0 + \delta$, we have
$$f(x_0) \leq f(x) < l + \varepsilon,$$
which implies $f(x_0) < l + \varepsilon$. Similarly, we will get $l - \varepsilon < f(x_0)$. Hence $|f(x_0) - l| < \varepsilon$. Since ε was an arbitrary positive number, we must have $f(x_0) = l$. This completes the proof of our statement. Note that if one assumes that $f(x)$ is decreasing, then one might have taken $g(x) = -f(x)$, which is an increasing function, and use the previous proof to obtain the desired conclusion. Of course, a direct proof based on similar ideas will work as well.

Solution 4.12

Without loss of generality, we may assume that $f(x)$ is increasing. Let $x_0 \in (a,b)$. Set

$$U(x_0) = \inf\{f(x); x \in (x_0, b)\} \quad \text{and} \quad L(x_0) = \sup\{f(x); x \in (a, x_0)\} \,.$$

Note that for any $x \in (x_0, b)$, we have $f(x_0) \leq f(x)$. Hence the set $\{f(x); x \in (x_0, b)\}$ is bounded below by $f(x_0)$ which implies the existence of $U(x_0)$ and forces the inequality $f(x_0) \leq U(x_0)$. Similarly for any $x \in (a, x_0)$, we have $f(x) \leq f(x_0)$. Hence the set $\{f(x); x \in (a, x_0)\}$ is bounded above by $f(x_0)$ which implies the existence of $L(x_0)$ and forces the inequality $L(x_0) \leq f(x_0)$. So we have

$$L(x_0) \leq f(x_0) \leq U(x_0) \,.$$

First we claim $\lim_{x \to x_0+} f(x) = U(x_0)$. Indeed, let $\varepsilon > 0$. Then by definition of $U(x_0)$, there exists $x^* \in (x_0, b)$ such that $U(x_0) \leq f(x^*) < U(x_0) + \varepsilon$. Set $\delta = x^* - x_0$. Then $\delta > 0$ since $x_0 < x^*$. Let $x \in (x_0, x_0 + \delta) = (x_0, x^*)$, then we have

$$U(x_0) \leq f(x) \leq f(x^*) < U(x_0) + \varepsilon \,.$$

In particular we have $|f(x) - U(x_0)| < \varepsilon$ which completes the proof of our claim. A similar proof will imply $\lim_{x \to x_0-} f(x) = L(x_0)$. For the last part of this problem similar ideas as the ones described above will show

$$\lim_{x \to a+} f(x) = U(a) \quad \text{and} \quad \lim_{x \to b-} f(x) = L(b) \,.$$

Note that $L(a)$ and $U(b)$ do not exist.

Solution 4.13

We have seen in the previous problems that for any $x_0 \in (a, b)$, then $\lim_{x \to x_0+} f(x)$ and $\lim_{x \to x_0-} f(x)$ exist. In particular, we have

$$\lim_{x \to x_0-} f(x) \leq f(x_0) \leq \lim_{x \to x_0+} f(x) \,.$$

In particular, $x_0 \in D \cap (a, b)$ if and only if

$$\lim_{x \to x_0-} f(x) < \lim_{x \to x_0+} f(x) \,.$$

So if $x \notin D$, then we must have

$$\lim_{x \to x_0-} f(x) = \lim_{x \to x_0+} f(x) = f(x_0) \,,$$

or that $\lim_{x \to x_0} f(x) = f(x_0)$. On the other hand, let $\varepsilon > 0$. Set

$$D_\varepsilon = \left\{ x_0 \in D \cap (a, b) \,,\, \lim_{x \to x_0+} f(x) - \lim_{x \to x_0-} f(x) > \varepsilon \right\} \,.$$

We claim that D_ε is finite. Indeed, let $x_i \in D_\varepsilon$, $i = 1, \ldots, n$, such that $x_1 < x_2 < \ldots < x_n$. Then it is easy to check that

$$f(a) \leq \lim_{x \to x_1-} f(x) < \lim_{x \to x_1+} f(x) \leq \ldots \leq \lim_{x \to x_n-} f(x) < \lim_{x \to x_n+} f(x) \leq f(b) \,.$$

This will then imply

$$\sum_{i=1}^{n} \lim_{x \to x_i+} f(x) - \lim_{x \to x_i-} f(x) \leq f(b) - f(a) \ .$$

Using the definition of D_ε we get $n\varepsilon < f(b) - f(a)$. Since the number n is bounded above, then D_ε must be finite. Finally note

$$D = \bigcup_{n \geq 1} D_{\frac{1}{n}} \ ,$$

which implies that D is a countable union of finite sets. Hence D is countable. Note that if $f : \mathbb{R} \to \mathbb{R}$ is monotone, then the set $D = \{x : x \in \mathbb{R} \text{ and } f \text{ does not have a limit at } x\}$ is countable. Indeed, We have

$$D = \bigcup_{n \geq 1} D \cap (-n, n) \ .$$

Since $D \cap (-n, n)$ is countable, then D is a countable union of countable sets. So D is countable.

Solution 4.14

The equation satisfied by $f(x)$ is usually known as a functional equation. First note that $f(nx) = nf(x)$ for any $n \in \mathbb{N}$. Indeed, we have $f(1 \cdot x) = 1 \cdot f(x)$. Assume that $f(nx) = nf(x)$. Then

$$f((n+1)x) = f(nx + x) = f(nx) + f(x) = nf(x) + f(x) = (n+1)f(x) \ .$$

By induction we get the desired identity. Let $r = \dfrac{p}{q} \in \mathbb{Q}$. Then we have $f(qrx) = f(px) = pf(x)$, and since $f(qrx) = qf(rx)$, we get $f(rx) = \dfrac{p}{q} f(x) = rf(x)$. In particular, we have $f(r) = rf(1)$. Let $x \in \mathbb{R}$. Then there exists a sequence of rational numbers $\{r_n\}$ such that $\lim\limits_{n \to \infty} r_n = x$. Since $f(x) = f(x - r_n + r_n) = f(x - r_n) + f(r_n) = f(x - r_n) + r_n f(1)$, we have $f(x) = f(x - r_n) + r_n f(1)$. Now if we take the limit as $n \to \infty$, by the existence of $\lim\limits_{x \to 0} f(x) = f(0)$, we have that $f(x) = f(0) + xf(1)$. Note that $f(x) = f(x + 0) = f(x) + f(0)$ for any $x \in \mathbb{R}$, so $f(0) = 0$. Therefore, we have

$$f(x) = xf(1) = mx \ .$$

Solution 4.15

1. $\lim\limits_{x \to \infty} \dfrac{x^2 + x}{x^2} = \lim\limits_{x \to \infty} \left(1 + \dfrac{1}{x}\right) = 1$ and therefore $x^2 + x \sim x^2$.

2. This follows from part (1).

3. We can see that $e^{\sqrt{\log x}} = o(x)$ because $\dfrac{e^{\sqrt{\log x}}}{x} = e^{\sqrt{\log x} - \log x}$. Now because $\lim\limits_{x \to \infty} \log x \to \infty$ and $\lim\limits_{x \to \infty} \dfrac{\sqrt{\log x}}{\log x} \to 0$, we have that for large x, $\sqrt{\log x} \leq \dfrac{\log x}{2}$. Hence, for similarly large x, this gives $\dfrac{e^{\sqrt{\log x}}}{x} \leq \dfrac{1}{x}[e^{\frac{\log x}{2}}] = \dfrac{1}{x}[\sqrt{x}] = \dfrac{1}{\sqrt{x}}$ and so $\lim\limits_{x \to \infty} \dfrac{1}{\sqrt{x}} = 0$. Thus we have showed that $e^{\sqrt{\log x}} = o(x)$.

4. First note that if we set $y = \frac{x^{\log x}}{e^x}$, then $\log y = \log x^{\log x} - x = (\log x)^2 - x$. Next, we claim that $\lim_{x \to \infty} \frac{(\log x)^2}{x} = 0$. This comes from applying L'Hôpital's rule twice. Namely:

$$\lim_{x \to \infty} \frac{(\log x)^2}{x} = \lim_{x \to \infty} \frac{2(\log x)\frac{1}{x}}{1} = \lim_{x \to \infty} \frac{2\log x}{x} = \lim_{x \to \infty} \frac{2/x}{1} = 0.$$

Therefore, there must be a constant $k > 0$ such that $x > k$ implies $\frac{(\log x)^2}{x} < \frac{1}{2}$, or equivalently, $(\log x)^2 < \frac{x}{2}$. Therefore, for $x > k$, we have that $\log y = (\log x)^2 - x < -\frac{x}{2}$, so that means $0 < y < e^{-x/2} = \frac{1}{e^{x/2}}$. But as $x \to \infty$, $\frac{1}{e^{x/2}} \to 0$ so $y \to 0$. That is, $\lim_{x \to \infty} y = \lim_{x \to \infty} \frac{x^{\log x}}{e^x} = 0$, which gives the desired result: $x^{\log x} = o(e^x)$.

Solution 4.16

Assume $g(x) > 0$ since $g(x) \to \infty$. Then since $f(x) = o(g(x))$ as $x \to \infty$, there is $k > 0$ such that $|f(x)| < \frac{g(x)}{2}$ for $x > k$. Therefore, $f(x) - g(x) < -\frac{g(x)}{2}$ for $x > k$ and $\lim_{x \to +\infty}(f(x) - g(x)) \to -\infty$. Now notice that $\frac{e^{f(x)}}{e^{g(x)}} = e^{f(x)-g(x)}$, and thus $e^{f(x)-g(x)} \to 0$ as $x \to +\infty$. Therefore $e^{f(x)} = o(e^{g(x)})$, as claimed.

Chapter 5

Continuity

The majority of my readers will be greatly disappointed to learn that by this commonplace observation the secret of continuity is to be revealed.

Julius Wilhelm Richard Dedekind (1831–1916)

- Let $f : D \to \mathbb{R}$ and let $c \in D$. We say that f is *continuous at* c if for every $\varepsilon > 0$ there exists a $\delta > 0$ such that $|f(x) - f(c)| < \varepsilon$ whenever $|x - c| < \delta$ and $x \in D$. If f is continuous at each point of a subset $K \subseteq D$, then f is said to be *continuous on* K. Moreover, if f is continuous on its domain D, then we simply say that f is *continuous*.

- Let D be a nonempty subset of $\mathbb{R}$ and $f : D \to \mathbb{R}$. We say that f is *uniformly continuous* on D if for every $\varepsilon > 0$ there exists a $\delta > 0$ such that $|x - c| < \delta$ and $x, c \in D$ imply $|f(x) - f(c)| < \varepsilon$. Notice that the δ in this definition depends on ε and f, but not on the point c or x. The geometric illustration of uniform continuity is shown in Figure 5.1.

- Let $f : A \subset \mathbb{R} \to \mathbb{R}$ be continuous and A is a closed and bounded subset of $\mathbb{R}$. Then f is uniformly continuous on A.

- *Intermediate Value Theorem*: If $f : [a, b] \to \mathbb{R}$ is continuous, and if L is a real number satisfying $f(a) < L < f(b)$ or $f(a) > L > f(b)$, then there is a point $c \in (a, b)$ where $f(c) = L$.

- A function f defined on an interval I is said to be *convex* if
$$f(\alpha x + (1 - \alpha)y) \leq \alpha f(x) + (1 - \alpha)f(y)$$
holds true for any x, y in I and $0 \leq \alpha \leq 1$.

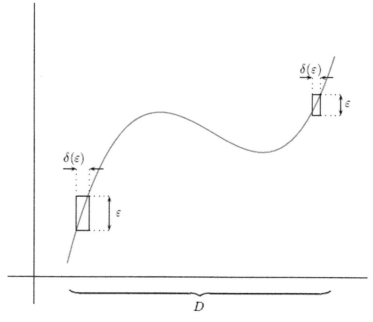

Figure 5.1

- We say f is a *periodic* function on $\mathbb{R}$, if there is a $T \in \mathbb{R}$ such that
$$f(x+T) = f(x)$$
for all $x \in \mathbb{R}$.

- For $f : \mathbb{R} \to \mathbb{R}$, $c \in \mathbb{R}$, and $\delta > 0$, define the *modulus of continuity* $w_f(c, \delta)$ of f by
$$w_f(c, \delta) = \sup\{|f(x) - f(c)| : x \in \mathbb{R},\ |x - c| < \delta\}.$$

Problem 5.1 Prove that if c is an isolated point in D, then f is automatically continuous at c.

Problem 5.2 Show that the value δ in the definition of continuity is not unique.

CHAPTER 5. CONTINUITY

Problem 5.3 Use the definition of continuity at a point to prove that

(a) $f(x) = 3x - 5$ is continuous at $x = 2$.

(b) $f(x) = x^2$ is continuous at $x = 3$.

(c) $f(x) = 1/x$ is continuous at $x = 1/2$.

Problem 5.4 Let $f(x) = x\sin(1/x)$ for $x \neq 0$ and $f(0) = 0$. Prove that f is continuous at $x = 0$.

Problem 5.5 Let $f : D \to \mathbb{R}$ and let $c \in D$. Show that the following conditions are equivalent.

(i) f is continuous at c.

(ii) If $\{x_n\}$ is a sequence in D such that $\{x_n\}$ converges to c, then
$$\lim_{n \to \infty} f(x_n) = f(c).$$

(This condition is called *sequential continuity*).

Problem 5.6 Let $f(x) = 1/x$ for $x \neq 0$ and $f(0) = c$ for some $c \in \mathbb{R}$. Show that $f(x)$ is not continuous at $x = 0$.

Problem 5.7 A function $f : \mathbb{R} \to \mathbb{R}$ is called *Lipschitz* with Lipschitz constant $\alpha > 0$ if
$$|f(x) - f(y)| \leq \alpha |x - y| \quad \text{for all } x, y \in \mathbb{R}.$$
Give two examples of Lipschitz functions. Moreover, prove that every Lipschitz function is continuous.

Problem 5.8 Find an example of a function that is discontinuous at every real number.

Problem 5.9 Prove the Dirichlet function $f : (0,1) \to \mathbb{R}$ defined as
$$f(x) = \begin{cases} 0 & \text{if } x \in \mathbb{R} \setminus \mathbb{Q} \\ \frac{1}{q} & \text{if } x \in \mathbb{Q} \text{ and } x = \frac{p}{q} \text{ in lowest terms} \end{cases}$$

(a) discontinuous at every rational number in $(0,1)$,

(b) continuous at each irrational number in $(0,1)$.

Problem 5.10 Let $f : [a, b] \to \mathbb{R}$ be continuous. Prove that $|f|$ is also continuous on $[a, b]$. Is the converse true? Namely, if $|f|$ is continuous on $[a, b]$, is f also continuous on $[a, b]$?

Problem 5.11 For $f : \mathbb{R} \to \mathbb{R}$, $c \in \mathbb{R}$, and $\delta > 0$, define the modulus of continuity $w_f(c, \delta)$ of f by
$$w_f(c, \delta) = \sup\{|f(x) - f(c)| : x \in \mathbb{R}, \ |x - c| < \delta\} \quad \text{and} \quad w_f(c) = \lim_{\delta \to 0^+} w_f(c, \delta).$$
Show that f is continuous at c if and only if $w_f(c) = 0$.

Problem 5.12 Find an example of a function f discontinuous on $\mathbb{Q}$ and another function g discontinuous at only one point, but $g \circ f$ is nowhere continuous.

Problem 5.13 Suppose $f : \mathbb{R} \to \mathbb{R}$ satisfies $f(x + y) = f(x) + f(y)$ for each $x, y \in \mathbb{R}$. Show that

(a) $f(nx) = nf(x)$ for all $x \in \mathbb{R}, n \in \mathbb{N}$;

(b) f is continuous at a single point if and only if f is continuous on $\mathbb{R}$;

(c) f is continuous if and only if $f(x) = mx$, for some $m \in \mathbb{R}$.

Problem 5.14 Let $f(x) = [x]$ be the greatest integer less than or equal to x and let $g(x) = x - [x]$. Sketch the graphs of f and g. Determine the points at which f and g are continuous.

Problem 5.15 If f, g are continuous functions defined on some subset $D \subseteq \mathbb{R}$, prove the $\max(f, g)$ and $\min(f, g)$ are continuous functions, where
$$\max(f, g)(x) = \max(f(x), g(x)) \quad \text{and} \quad \min(f, g)(x) = \min(f(x), g(x)).$$

Problem 5.16 Let I be a closed and bounded interval in $\mathbb{R}$. Suppose $f : I \to \mathbb{R}$ is continuous. Show that $f(I)$ is a closed and bounded interval in $\mathbb{R}$.

Problem 5.17 Let I be a closed and bounded subset of $\mathbb{R}$ and $f : I \to \mathbb{R}$ be continuous. Prove that f assumes maximum and minimum values on I, i.e., there exist $x_1, x_2 \in I$ such that
$$f(x_1) \leq f(x) \leq f(x_2)$$
for all $x \in I$.

CHAPTER 5. CONTINUITY

Problem 5.18 Let $f, g : [0,1] \to [0, \infty)$ be continuous functions satisfying
$$\sup_{0 \leq x \leq 1} f(x) = \sup_{0 \leq x \leq 1} g(x).$$
Prove that there exists $x_0 \in [0,1]$ such that $f(x_0) = g(x_0)$.

Problem 5.19 Prove that if $f : \mathbb{R} \to \mathbb{R}$ is continuous and periodic, then it attains its supremum and infimum.

Problem 5.20 Let $f : [0,1] \to [0,1]$ be continuous. Prove that f has a fixed point. That is, show that there is a point $x_0 \in [0,1]$ such that $f(x_0) = x_0$.

Problem 5.21 Prove that $f(x) = x^2$ is uniformly continuous on $[0,1]$ but not uniformly continuous on $\mathbb{R}$.

Problem 5.22 Prove that if I is a closed and bounded interval in $\mathbb{R}$ and if $f : I \to \mathbb{R}$ is continuous, then f is uniformly continuous on I.

Problem 5.23 Show that $f : \mathbb{R} \to \mathbb{R}$ is not uniformly continuous if and only if there exists an $\varepsilon > 0$ and sequences x_n and y_n such that $|x_n - y_n| < \frac{1}{n}$ and $|f(x_n) - f(y_n)| \geq \varepsilon$.

Problem 5.24 Determine which of the following functions are uniformly continuous:

(a) $f(x) = \ln x$ on $(0, 1)$.

(b) $f(x) = x \sin x$ on $[0, \infty)$.

(c) $f(x) = \sqrt{x}$ on $[0, \infty)$.

(d) $f(x) = \frac{1}{x^2+1}$ on $(-\infty, \infty)$.

(e) $f(x) = e^x$ on $[0, \infty)$.

Problem 5.25 Suppose $I \subseteq \mathbb{R}$ and $f : I \to \mathbb{R}$ is uniformly continuous. Prove that if (x_n) is a Cauchy sequence in I, then $(f(x_n))$ is Cauchy too. (Recall that a sequence of points $x_n \in \mathbb{R}$ is said to be Cauchy if for every $\varepsilon > 0$ there exists an $N \in \mathbb{N}$ such that $n, m \geq N$ implies that $|x_n - x_m| < \varepsilon$.)

Problem 5.26 Find a continuous function $f : I \to \mathbb{R}$ and a Cauchy sequence (x_n) such that $(f(x_n))$ is not Cauchy.

Problem 5.27 Show that if a function $f : (a,b) \to \mathbb{R}$ is uniformly continuous, then we can extend it to a function $\tilde{f}$ that is also uniformly continuous on $[a,b]$.

Problem 5.28 Show that any function continuous and periodic on $\mathbb{R}$ must be uniformly continuous.

Problem 5.29 Prove that if a function f defined on an open interval $I \subseteq \mathbb{R}$ is convex, then f is continuous. Must a convex function on an arbitrary interval be continuous?

Problem 5.30 A function $\phi : [a,b] \to \mathbb{R}$ is called a step function if there is a partition $a = x_0 < x_1 < \ldots < x_n = b$ such that $\phi(x)$ is constant on (x_{i-1}, x_i) for $i = 1, 2, 3, \ldots, n$. Prove that if $f : [a,b] \to \mathbb{R}$ is continuous, then f can be uniformly approximated by step functions. Note that we call a mapping g on $[a,b]$ a uniform approximation of the mapping f if there is some number $\varepsilon > 0$ such that $|f(x) - g(x)| \leq \varepsilon$.

Problem 5.31 Show that there exists a continuous function $F : [0,1] \to \mathbb{R}$ whose derivative exists and equals zero almost everywhere but which is not constant.

Problem 5.32 Throughout $\mathbb{R}^{+\bullet} = \{x \in \mathbb{R} : x > 0\}$.

Let $f : \mathbb{R}^{+\bullet} \to \mathbb{R}^{+\bullet}$ be a function, and let $a \in \mathbb{R}^{+\bullet}$. Then f is feebly continuous at a if there is a sequence $(r_n) \searrow 0$ such that $f(a + r_n) \to f(a)$ as $n \to \infty$.

Let (f_j) be a sequence of functions from $\mathbb{R}^{+\bullet}$ to $\mathbb{R}^{+\bullet}$, and let $a \in \mathbb{R}^{+\bullet}$. Then (f_j) is feebly continuous at a if there is a sequence $(r_n) \searrow$ such that $f_j(a + r_n) \to f_j(a)$ as $n \to \infty$ for each $j \in \mathbb{N}$.

Show that each sequence of functions from $\mathbb{R}^{+\bullet}$ to $\mathbb{R}^{+\bullet}$ is feebly continuous at all but countably many points of $\mathbb{R}^{+\bullet}$.
Notice that the solution of the above question leads to the fact that each function from $\mathbb{R}^+$ to $\mathbb{R}^+$ is feebly continuous at all but countably many points of $\mathbb{R}^+$.
This question and its solution are provided by Professor H. G. Dales from University of Leeds.

Problem 5.33 (**Banach Contraction Mapping Theorem**) Let f be a function defined on all of $\mathbb{R}$, and assume that there is a constant k such that $0 < k < 1$ and
$$|f(x) - f(y)| \leq k|x - y|$$
for all $x, y \in \mathbb{R}$. (In this case we call f a contraction mapping.)

a) Show that f is continuous on $\mathbb{R}$.

b) Pick some point $x_0 \in \mathbb{R}$ and construct the sequence $x_{n+1} = f(x_n)$ more precisely
$$(x_0, f(x_0), f(f(x_0)), \ldots).$$
Show that the resulting sequence (x_n) is a Cauchy sequence and converges to some point x^* in $\mathbb{R}$.

c) Show that x^* is a fixed point of f (i.e., $f(x^*) = x^*$).

d) Show that f has a unique fixed point.

Solutions

Solution 5.1

Suppose c is an isolated point of D. Then there exist $\delta > 0$ such that the δ-neighborhood of c contains no other point in D (i.e., if $|x-c| < \delta$ and $x \in D$, then $x = c$). Thus, whenever $|x-c| < \delta$ and $x \in D$, we have $|f(x) - f(c)| = 0 < \varepsilon$ for all $\varepsilon > 0$.

Solution 5.2

Once one value of δ is found that fulfills the requirements of the definition, any smaller positive value for δ will also fulfill the requirements. Suppose f is continuous at c. Then
$$|f(x) - f(c)| < \varepsilon \quad \text{if} \quad 0 < |x-c| < \delta.$$
Choosing any δ_1 such that $0 < \delta_1 < \delta$, we observe that $0 < |x-c| < \delta_1 < \delta$ will also give us $|f(x) - f(c)| < \varepsilon$.

Solution 5.3

(a) We must show that given any positive ε, we can find a positive δ such that
$$|f(x) - f(c)| = |(3x-5) - 1| < \varepsilon \quad \text{if } x \text{ satisfies} \quad 0 < |x-2| < \delta.$$
This simplifies to
$$|3x - 6| < \varepsilon \quad \text{if} \quad 0 < |x-2| < \delta,$$
and further simplifies to
$$3|x-2| < \varepsilon \quad \text{if} \quad 0 < |x-2| < \delta.$$
A choice of δ that makes this last "if statement" true for any ε is $\delta = \varepsilon/3$.

(b) Given any positive $\varepsilon > 0$, we must find a positive δ such that
$$|x^2 - 9| < \varepsilon \quad \text{if} \quad 0 < |x-3| < \delta,$$
which we may factor to
$$|x-3||x+3| < \varepsilon \quad \text{if} \quad 0 < |x-3| < \delta.$$
Now, if there were some constant C such that $|x+3| < C$, we would then have $|x-3||x+3| < C|x-3|$, and thus
$$|x-3||x+3| < C|x-3| < \varepsilon \quad \text{if} \quad 0 < |x-3| < \delta$$
for the choice $\delta = \varepsilon/C$. But what value of C would this work for? By Problem 5.2, we know that if we find a δ that works, then any smaller positive δ can be used in its place. This allows us to assume $\delta \leq 1$, because if $\delta > 1$ we may use the value $\delta = 1$ instead. Therefore,
$$0 < |x-3| < \delta \leq 1 \quad \Rightarrow \quad |x+3| < 7.$$
Thus, we let $C = 7$. Given $\varepsilon > 0$, we then may choose $\delta = \varepsilon/7$, provided $\varepsilon/7 \leq 1$. More precisely, choose $\delta = \min(\varepsilon/7, 1)$.

(c) For all $\varepsilon > 0$ we must find a $\delta > 0$ such that

$$\left|\frac{1}{x} - 2\right| < \varepsilon \quad \text{if} \quad 0 < \left|x - \frac{1}{2}\right| < \delta.$$

Rewriting this as

$$\left|\frac{1}{x} - 2\right| = \left|\frac{2}{x}\left(\frac{1}{2} - x\right)\right| = \left|\frac{2}{x}\right|\left|\frac{1}{2} - x\right|,$$

we wish to find C such that $|2/x| < C$. However, the graph of $|2/x|$ makes it clear that $\delta \geq 1/2$ will not produce such an upper bound C. Therefore, arbitrary restriction on δ must be smaller than $1/2$. We then might restrict $\delta \leq 1/4$ to give

$$\left|\frac{2}{x}\right|\left|x - \frac{1}{2}\right| < \varepsilon \quad \text{if} \quad 0 < \left|x - \frac{1}{2}\right| < \delta \leq \frac{1}{4},$$

which yields $8|x - 1/2| < \varepsilon$, so the choice $\delta = \varepsilon/8$ is appropriate. Therefore, given $\varepsilon > 0$, we choose $\delta = \varepsilon/8$ provided that this does not violate the $\delta \leq 1/4$. Thus, our more general choice is $\delta = \min(1/4, \varepsilon/8)$.

Solution 5.4

Since $|f(x) - f(0)| = |x\sin(1/x)| \leq |x|$ for all x, given $\varepsilon > 0$ we may choose $\delta = \varepsilon$. We then have $|x - 0| < \delta$ implies $|f(x) - f(0)| \leq |x| < \delta = \varepsilon$.

Solution 5.5

1. Let us first prove (i)$\Rightarrow$(ii). Suppose that $\{x_n\}$ is a sequence with $\lim_{n\to\infty} x_n = c$. We need to show that $\lim_{n\to\infty} f(x_n) = f(c)$. Let $\varepsilon > 0$. We must find an integer N so that $n \geq N$ implies $|f(x_n) - f(c)| < \varepsilon$. To do this, choose $\delta > 0$ so that $|x - c| < \delta$ implies $|f(x) - f(c)| < \varepsilon$. The existence of such a δ is guaranteed by the continuity of f. Since $\lim_{n\to\infty} x_n = c$, there exists $N \geq 1$ such that $n \geq N$ implies $|x_n - c| < \delta$. Hence for any $n \geq N$, we have $|f(x_n) - f(c)| < \varepsilon$. This yields the desired conclusion.

2. Next we show (ii) $\Rightarrow$ (i). Suppose, to the contrary, that f is sequentially continuous but discontinuous at c. Then there is a neighborhood V of $f(c)$ such that no neighborhood U of c satisfies $f(U) \subseteq V$. Set $U_n = B(c, 1/n) = \{x \in D : |c - x| < 1/n\}$, for $n \geq 1$. In particular, we have that $f(U_n) \cap V^c \neq \emptyset$, for any $n \geq 1$. Hence for $n \geq 1$, choose some $x_n \in D$ with $|x - c| < 1/n$ and $f(x_n) \notin V$. By construction, $\{x_n\}$ converges to c, but $\{f(x_n)\}$ does not converge to $f(c)$. This contradicts the sequential continuity of f.

Solution 5.6

Let the sequence $\{x_n\}$ be defined by $x_n = 1/n$. We then note that

$$\lim_{n\to\infty} \frac{1}{n} = 0, \quad \text{but} \quad \lim_{n\to\infty} f\left(\frac{1}{n}\right) = +\infty.$$

Since the sequence $\{f(x_n)\}$ is not convergent, there is no way to define f at $x = 0$ to make f continuous at $x = 0$.

$\boxed{\text{Solution 5.7}}$

1. Any constant function $f(x) = k$, for some $k \in \mathbb{R}$.

2. The identity function $f(x) = x$.

To prove the assertion that every Lipschitz function is continuous, given $c \in \mathbb{R}$ and $\varepsilon > 0$, set $\delta = \varepsilon/\alpha$. Then $|f(x) - f(c)| < \varepsilon$ for all $x \in \mathbb{R}$ such that $|x - c| < \delta$. Note that in this case δ is independent of $c \in \mathbb{R}$.

$\boxed{\text{Solution 5.8}}$

Define $f : \mathbb{R} \to \mathbb{R}$ by

$$f(x) = \begin{cases} 1 & \text{if } x \in \mathbb{Q}, \\ 0 & \text{if } x \in \mathbb{R}\setminus\mathbb{Q}. \end{cases}$$

Let $c \in \mathbb{R}$. Assume $f(x)$ is continuous at c. For $\varepsilon = \frac{1}{2}$, there exists $\delta > 0$ such that $|f(x) - f(c)| < \varepsilon$ whenever $|x - c| < \delta$. The interval $(c - \delta, c + \delta)$ contains rational and irrational points. Let $r \in (c - \delta, c + \delta) \cap \mathbb{Q}$ and $r^* \in (c - \delta, c + \delta)\setminus\mathbb{Q}$. Then

$$|f(r) - f(c)| = |1 - c| < \varepsilon \text{ and } |f(r^*) - f(c)| = |0 - c| < \varepsilon.$$

This will force the inequality

$$1 = |1 - 0| \leq |1 - c| + |c - 0| < 2\varepsilon = 1.$$

The generated contradiction implies that $f(x)$ is not continuous at c. Alternatively, we can use sequential continuity. Indeed, note that for any $c \in \mathbb{R}$, there exist two sequences $\{r_n\} \in \mathbb{Q}$ and $\{r_n^*\} \in \mathbb{R}\setminus\mathbb{Q}$ which converge to c. Since $\{f(r_n)\}$ converges to 1 and $\{f(r_n^*)\}$ converges to 0, $f(x)$ is not sequentially continuous at c. So $f(x)$ is not continuous at c.

$\boxed{\text{Solution 5.9}}$

In Problem 4.3, we showed that $\lim_{x\to a} f(x) = 0$ for any $a \in (0, 1)$. Therefore

(a) if $a \in \mathbb{R}\setminus\mathbb{Q}$, then $f(a) = 0$ which implies $\lim_{x\to a} f(x) = f(a)$, i.e., $f(x)$ is continuous at a;

(b) if $a \in \mathbb{Q} \cap (0, 1)$, then $f(a) \neq 0$ which implies $\lim_{x\to a} f(x) \neq f(a)$, i.e., $f(x)$ is not continuous at a.

CHAPTER 5. CONTINUITY

Solution 5.10

The key behind the proof of this problem is the inequality

$$\Big||x| - |y|\Big| \leq |x - y|$$

for any $x, y \in \mathbb{R}$. Indeed, since f is continuous, given $x_0 \in [a, b]$ and $\varepsilon > 0$ we may find $\delta > 0$ such that if $x \in [a, b]$ and $0 < |x - x_0| < \delta$, then $|f(x) - f(x_0)| < \varepsilon$. The inequality

$$||f(x)| - |f(x_0)|| \leq |f(x) - f(x_0)| < \varepsilon$$

whenever $0 < |x - x_0| < \delta$. This implies the continuity of $|f|$ at x_0. The proposed converse of this statement is not true, however. For example, the function

$$f(x) = \begin{cases} 1 & \text{if } x \in \mathbb{Q} \cap [a, b], \\ -1 & \text{if } x \in [a, b] \backslash \mathbb{Q} \end{cases}$$

is discontinuous at each point in $[a, b]$, while $|f|$ is a constant function and therefore continuous on $[a, b]$.

Solution 5.11

1 First assume f is continuous at c. Then given any $\varepsilon > 0$, there is a $\delta_0 > 0$ such that

$$|x - c| < \delta_0 \text{ implies } |f(x) - f(c)| < \varepsilon/2.$$

Notice also that $w_f(c, \delta_1) \leq w_f(c, \delta_2)$ whenever $0 < \delta_1 < \delta_2$. Hence, if $0 < \delta < \delta_0$, then $w_f(c, \delta) \leq w_f(c, \delta_0) < \varepsilon$, and consequently $\lim_{\delta \to 0^+} w_f(c, \delta) = 0$.

2 Conversely assume that $w_f(c) = 0 = \lim_{\delta \to 0^+} w_f(c, \delta)$. Then given $\varepsilon > 0$, there exists some δ_0 such that $w_f(c, \delta) < \varepsilon$ if $\delta < \delta_0$. Consequently, if $|x - c| < \delta < \delta_0$, then $|f(x) - f(c)| < \varepsilon$.

Solution 5.12

Let f be the Dirichlet function defined in Problem 5.9 (but extended from $(0, 1)$ to all of $\mathbb{R}$) and set

$$g(x) = \begin{cases} 1 & \text{if } x \neq 0, \\ 0 & \text{if } x = 0. \end{cases}$$

We then have that

$$(g \circ f)(x) = \begin{cases} 1 & \text{if } x \in \mathbb{Q}, \\ 0 & \text{if } x \notin \mathbb{Q} \end{cases}$$

which is shown to be discontinuous at all points of $\mathbb{R}$ in Problem 5.8.

Solution 5.13

We begin by making the observation that $f(0) = f(0+0) = f(0) + f(0)$ which implies $f(0) = 0$. Hence
$$0 = f(0) = f(a + (-a)) = f(a) + f(-a)$$
which implies $f(-a) = -f(a)$. Since (a) is straightforward, we will only prove (b) and (c).

(b) Suppose f is continuous at a single point c. Then given $\varepsilon > 0$ there exists some $\delta > 0$ such that $|x - c| < \delta$ implies $|f(x) - f(c)| < \varepsilon$. Let d be any point of $\mathbb{R}$ and let $|x - d| < \delta$. Then $|x - d + c - c| < \delta$ implies $|f(x - d + c) - f(c)| < \varepsilon$. Using the given property for f, we obtain $|f(x - d) + f(c) - f(c)| = |f(x) - f(d)| < \varepsilon$. It follows that f is continuous at any point of $\mathbb{R}$. The converse is clear from the definition of continuity.

(c) Assume $f(x)$ is continuous and set $m = f(1)$. Then we know that $f(n) = nf(1) = mn$ for any $n \in \mathbb{Z}$. If $x \in \mathbb{Q}$, then $x = p/q$ for some $(p, q) \in \mathbb{N} \times \mathbb{Z}$, with $q \neq 0$. We have $f(qx) = f(p) = mp$ and $f(qx) = qf(x)$ which implies $f(x) = \dfrac{p}{q}m = mx$. Since $f(x)$ is continuous and $\mathbb{Q}$ is dense in $\mathbb{R}$, then we get $f(x) = mx$ for all $x \in \mathbb{R}$. The converse is obvious since $f(x) = mx$ is continuous.

Solution 5.14

The graphs of $f(x)$ and $g(x)$ are shown in the following figure.

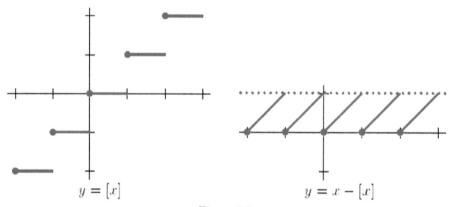

Figure 5.2

Notice that if $n \in \mathbb{N}$ and $n \leq x < n+1$, then $f(x) = [x] = n$ and $g(x) = x - [x] = x - n$. Therefore, $0 \leq g < 1$. It is clear that f and g are continuous on $(n, n+1)$ for every $n \in \mathbb{Z}$. But since
$$\lim_{x \to n+} f(x) = n = f(n) \text{ and } \lim_{x \to n+} g(x) = n - n = 0 = g(n)$$
and
$$\lim_{x \to n-} f(x) = n - 1 \neq f(n) \text{ and } \lim_{x \to n+} g(x) = n - (n-1) = 1 \neq g(n)$$
for all $n \in \mathbb{Z}$, we conclude that $f(x)$ and $g(x)$ are not continuous at any point in $\mathbb{Z}$.

Solution 5.15

The key behind the proof of these statements are the identities

$$\max(x,y) = \frac{1}{2}(x+y+|x-y|) \text{ and } \min(x,y) = \frac{1}{2}(x+y-|x-y|)$$

for all $x, y \in \mathbb{R}$. Since $f \pm g$ and $|f - g|$ are continuous, we conclude that $\max(f, g)$ and $\min(f, g)$ are continuous.

Solution 5.16

Suppose $f(I)$ is not bounded. Then for each $n \in \mathbb{N}$ there exists a point $x_n \in I$ such that $|f(x_n)| > n$. Since I is closed and bounded, by the Bolzano–Weierstrass Theorem, the sequence (x_n) has a convergent subsequence (x_{n_k}) converging to some $c \in I$. Since f is continuous at c, then $f(x_{n_k}) \to f(c)$. In particular, $f(x_{n_k})$ must be bounded. But this contradicts the fact that $|f(x_{n_k})| > n_k \geq k$ for all $k \in \mathbb{N}$. Therefore $f(I)$ is bounded.

To show that $f(I)$ is closed, take a sequence (y_n) in $f(I)$ with $\lim_{n \to \infty} y_n = y$. We want to show that $y \in f(I)$. Since $y_n \in f(I)$, there exists $x_n \in I$ such that $f(x_n) = y_n$ for all n. Since I is closed and bounded, it follows that there is a subsequence (x_{n_k}) converging to some point c, and we have

$$f(c) = \lim_{k \to \infty} f(x_{n_k}) = \lim_{k \to \infty} y_{n_k} = y.$$

Thus $f(c) = y$ which implies $y \in f(I)$.

Solution 5.17

By the previous question, we know $f(I)$ is closed and bounded. Since any nonempty closed and bounded subset of $\mathbb{R}$ has a maximum and a minimum, $f(I)$ has a minimum y_1 and a maximum y_2. But $y_1, y_2 \in f(I)$ means that there exist $x_1, x_2 \in I$ such that $y_1 = f(x_1)$ and $y_2 = f(x_2)$. It follows that $f(x_1) \leq f(x) \leq f(x_2)$ for all $x \in I$.

Solution 5.18

Let $M = \sup_{0 \leq x \leq 1} f(x) = \sup_{0 \leq x \leq 1} g(x)$. Since f and g are continuous on $[0, 1]$ and $[0, 1]$ is closed and bounded, there exist $c, d \in [0, 1]$ such that $f(c) = g(d) = M$. Without loss of generality assume $c < d$. Note that if $c = d$, we have nothing to prove. Now define the function $h(x) = f(x) - g(x)$. Then we have that

$$h(c) = f(c) - g(c) = M - g(c) \geq 0$$

and

$$h(d) = f(d) - g(c) = f(d) - M \leq 0.$$

Since h is continuous, from the Intermediate Value Theorem it has a zero x_0 in the interval $[c, d]$. So $h(x_0) = f(x_0) - g(x_0) = 0$.

Solution 5.19

Let $T > 0$ be a period of f. Since f is continuous on $\mathbb{R}$, it is continuous on $[0, T]$, i.e., there exist $c, d \in [0, T]$ such that $f(c) = \inf_{x \in [0,T]} f(x)$ and $f(d) = \sup_{x \in [0,T]} f(x)$. Since f is periodic, then we have

$$f(c) = \inf_{x \in [0,T]} f(x) = \inf_{x \in \mathbb{R}} f(x) \quad \text{and} \quad f(d) = \sup_{x \in [0,T]} f(x) = \sup_{x \in \mathbb{R}} f(x).$$

Solution 5.20

Assume $f(0) > 0$ and $f(1) < 1$, otherwise we have nothing to prove. Set $g(x) = f(x) - x$. Then g is continuous, and we have that $g(0) = f(0) - 0 > 0$ and $g(1) = f(1) - 1 < 0$. Hence, by the intermediate value theorem, g must equal 0 at some point $x_0 \in (0, 1)$, i.e., $g(x_0) = f(x_0) - x_0 = 0$.

Solution 5.21

To show that $f(x)$ is uniformly continuous on $[0, 1]$, given $\varepsilon > 0$ we must choose $\delta = \delta(\varepsilon)$ so that

$$|f(x) - f(c)| = |x^2 - c^2| < \varepsilon \quad \text{if} \quad |x - c| < \delta \quad \text{for any} \quad x, c \in [0, 1].$$

If we set $\delta = \varepsilon/3$, then

$$|f(x) - f(c)| = |x^2 - c^2| = |x - c||x + c| \leq |x - c|(1 + 1) = 2|x - c| < \frac{2\varepsilon}{3} < \varepsilon.$$

Next, assume that $f(x) = x^2$ is uniformly continuous on $\mathbb{R}$. Then for all $\varepsilon > 0$ there exists $\delta = \delta(\varepsilon) > 0$ such that

$$|x - c| < \delta \quad \Rightarrow \quad |f(x) - f(c)| < 1 \quad \text{for all} \quad x, c \in \mathbb{R}.$$

By the Archimedean Property, choose $n \in \mathbb{N}$ sufficiently large such that $n > 1/\delta$. Set $c = n$ and $x = n + \delta/2$. Since $|x - c| < \delta$, then

$$1 < n\delta < n\delta + \frac{\delta^2}{4} = n^2 + n\delta + \frac{\delta^2}{4} - n^2 = x^2 - c^2 = |f(x) - f(c)| < 1.$$

This contradiction proves that f is not uniformly continuous on $\mathbb{R}$.

Solution 5.22

Suppose to the contrary that $f : I \to \mathbb{R}$ is continuous but not uniformly continuous on I. Then there exists $\varepsilon_0 > 0$ such that for any $\delta > 0$, there exist $x_\delta, y_\delta \in I$ such that $|x_\delta - y_\delta| < \delta$ and $|f(x_\delta) - f(y_\delta)| \geq \varepsilon_0$. In particular, there must exist points $x_n, y_n \in I$ such that

$$|x_n - y_n| < 1/n \quad \text{and} \quad |f(x_n) - f(y_n)| \geq \varepsilon_0$$

for all $n \geq 1$. Since the sequence (x_n) is in I, which is closed and bounded, the Bolzano–Weierstrass Theorem gives that (x_n) has a subsequence (x_{n_k}) that converges to some $x \in I$ as $k \to \infty$. Similarly, some subsequence of the corresponding sequence (y_{n_k}), which we shall denote $(y_{n_{k_j}})$, converges to some $y \in I$ as $j \to \infty$. Since $x_{n_{k_j}} \to x$ as $j \to \infty$ and f is continuous, we must have $f(x_{n_{k_j}}) \to f(x)$, and similarly $f(y_{n_{k_j}}) \to f(y)$. From our assumptions it follows that $|f(x) - f(y)| \geq \varepsilon_0$, so

$f(x) \neq f(y)$. However, $|x_n - y_n| < 1/n$ for all $n \in \mathbb{N}$, so $\lim_{n \to \infty} x_n - y_n = 0 = x - y$. So we must have $x = y$, and thus $f(x) = f(y)$, a contradiction. It follows that $f(x)$ is uniformly continuous on I.

Solution 5.23

$\Rightarrow$: Assume f is not uniformly continuous, then there is an $\varepsilon > 0$ for which no $\delta > 0$ will work in the definition of uniform continuity. In particular, $\delta = \frac{1}{n}$ will not work. Therefore there must be a pair of numbers x_n and y_n such that

$$|x_n - y_n| < \frac{1}{n} \quad \text{but} \quad |f(x_n) - f(y_n)| \geq \varepsilon.$$

These x_n and y_n form required sequences.

$:\Leftarrow$ By the Archimedean principle we can select an integer n so that $0 < \frac{1}{n} < \delta$. For the corresponding x_n and y_n one will have $|f(x_n) - f(y_n)| \geq \varepsilon$ even though $|x_n - y_n| < \frac{1}{n} < \delta$. Since this can be done for every $\delta > 0$, f cannot be uniformly continuous on $\mathbb{R}$.

Solution 5.24

(a) Not uniformly continuous on $(0,1)$, because $|f(e^{-n}) - f(e^{-(n+1)})| = 1$ while $e^{-n} \to 0$ as $n \to \infty$.

(b) Not uniformly continuous on $[0, \infty)$, because $|f(2n\pi) - f(2n\pi + 1/n)| \to 2\pi$ as $n \to \infty$.

(c) Uniformly continuous on $[0, \infty)$. We note that

$$|f(x) - f(c)| = |\sqrt{x} - \sqrt{c}| \leq \sqrt{|x - c|} \quad \text{for} \quad x, c \in [0, \infty).$$

Therefore, given $\varepsilon > 0$, choose $\delta = \varepsilon^2$, then $|x - c| < \delta$ implies $|\sqrt{x} - \sqrt{c}| < \varepsilon$.

(d) Since $f(x)$ is even and continuous on $\mathbb{R}$, it is enough to show that $f(x)$ is uniformly continuous on $[1, \infty)$ since any continuous function on $[-1, 1]$ is uniformly continuous. On the other hand, we have

$$|f(x) - f(c)| = \left| \frac{1}{x^2 + 1} - \frac{1}{c^2 + 1} \right| \leq |x - c|$$

for any $x, c \in [1, \infty)$. Hence $f(x)$ is Lipschitz on $[1, \infty)$ which implies that $f(x)$ is uniformly continuous on $[1, \infty)$.

(e) Not uniformly continuous on $[0, \infty)$ because $|f(\ln n) - f(\ln(n+1))| = 1$ while

$$|\ln n - \ln(n+1)| = \ln\left(\frac{n}{n+1}\right) = \ln\left(1 + \frac{1}{n}\right) \to 0 \quad \text{as } n \to \infty.$$

Solution 5.25

Let $\varepsilon > 0$ and choose $\delta > 0$ such that

$$|x - c| < \delta \quad \text{and} \quad x, c \in I \quad \text{imply } |f(x) - f(c)| < \varepsilon.$$

Since (x_n) is Cauchy, choose $N \in \mathbb{N}$ such that $n, m \geq N$ implies $|x_n - x_m| < \delta$. Then $n, m \geq N$ implies $|f(x_n) - f(x_m)| < \varepsilon$. Since ε can be chosen arbitrarily, we have that $(f(x_n))$ is Cauchy.

Solution 5.26

Consider $f : (0, 1) \to \mathbb{R}$ defined by $f(x) = 1/x$. Then f is continuous on $(0, 1)$. If we set $(x_n) = (1/n)$, we have $x_n \to 0$ and therefore is Cauchy, but $(f(x_n))$ is divergent and thus is not Cauchy.

Solution 5.27

Suppose that f is uniformly continuous on (a, b) and let (x_n) be a sequence in (a, b) that converges to a. We claim that $\lim_{x \to a} f(x)$ exists. Since (x_n) is a Cauchy sequence, Problem 5.25 gives that $(f(x_n))$ is a Cauchy sequence in $\mathbb{R}$. Hence $(f(x_n))$ is convergent. The sequential characterization of the limit of a function (see Problem 4.8) implies $\lim_{x \to a} f(x)$ exists. A similar argument will show that $\lim_{x \to b} f(x) = q$ for some $q \in \mathbb{R}$. We may now define the extended function $\widetilde{f} : [a, b] \to \mathbb{R}$ by

$$\widetilde{f}(x) = \begin{cases} f(x) & \text{if } a < x < b, \\ p & \text{if } x = a, \\ q & \text{if } x = b. \end{cases}$$

Notice that $\widetilde{f}$ is continuous on $[a, b]$. Since $[a, b]$ is a closed and bounded interval, $\widetilde{f}$ is uniformly continuous on $[a, b]$ by Problem 5.22.

Solution 5.28

Let $T > 0$ be a period of f. Problem 5.22 implies that f is uniformly continuous on any interval $[a, b]$. Let us prove that f is in fact uniformly continuous on $\mathbb{R}$. Let $\varepsilon > 0$. Then there exists $\delta > 0$ such that for any $x, y \in [-T, 2T]$ such that $|x - y| < \delta$ we have $|f(x) - f(y)| < \varepsilon$. Without loss of generality we may assume that $\delta < T$. Now let $x, y \in \mathbb{R}$ such that $|x - y| < \delta$. There exists $n \in \mathbb{Z}$ such that $nT \leq x < (n+1)T$. Then $x - nT \in [0, T]$ and $y - nT \in [-T, 2T]$. Since $|(x - nT) - (y - nT)| = |x - y| < \delta$,

$$|f(x - nT) - f(y - nT)| = |f(x) - f(y)| < \varepsilon,$$

which completes the proof of our claim, i.e., f is uniformly continuous on $\mathbb{R}$.

Solution 5.29

Let $I = (a, b)$ be an open interval and assume f is convex on (a, b). Let $c \in (a, b)$. Suppose $a < s < c < d < t < b$. From the geometric interpretation of convexity, we know that the point $(c, f(c))$ lies below the line through the points $(s, f(s))$ and $(d, f(d))$. That is,

$$f(c) \leq f(s) + \frac{f(d) - f(s)}{d - s}(c - s).$$

Now consider the line through the points $(c, f(c))$ and $(t, f(t))$. Again from convexity we have

$$f(d) \leq f(c) + \frac{f(t) - f(c)}{t - c}(d - c).$$

CHAPTER 5. CONTINUITY   93

From the above inequalities we can obtain

$$f(s) + \frac{f(c) - f(s)}{c - s}(d - s) \leq f(d) \leq f(c) + \frac{f(t) - f(c)}{t - c}(d - c).$$

Since

$$\lim_{d \to c} f(s) + \frac{f(c) - f(s)}{c - s}(d - s) = f(c) \text{ and } \lim_{d \to c} f(c) + \frac{f(t) - f(c)}{t - c}(d - c) = f(c),$$

the Squeeze Theorem (Problem 4.9) implies $\lim_{d \to c} f(d) = f(c)$. Thus, f is continuous at an arbitrary point $c \in (a, b)$. It follows that f is continuous on (a, b).

If the interval is not open, then the convex function f need not be continuous, as illustrated in the following example:

$$f(x) = \begin{cases} x^2 & \text{if } x \in [0, 1), \\ 3 & \text{if } x = 1. \end{cases}$$

Solution 5.30

Since f is continuous on a closed and bounded interval, f is uniformly continuous. Hence, for all $\varepsilon > 0$ there exists a $\delta > 0$ such that $|x - y| < \delta$ implies $|f(x) - f(y)| < \varepsilon$. Choose $\delta' < \delta$ such that $n = (b - a)/\delta' \in \mathbb{N}$ and divide $[a, b]$ into n equal-length intervals such that $a = x_0 < x_1 < \ldots < x_n = b$. Clearly we have $x_{i+1} - x_i = \delta'$, for any i. For any $x \in [a, b]$, there exists a unique i such that $x \in [x_i, x_{i+1}]$. Define $\phi(x) = f(x_i)$ for $x_i \leq x < x_{i+1}$. Since $|x_i - x| \leq \delta' < \delta$, we get

$$|f(x_i) - f(x)| = |\phi(x) - f(x)| < \varepsilon,$$

which proves our claim.

Solution 5.31

We define the Cantor function (Devil's Staircase function) as follows:

$$F(x) = \begin{cases} \frac{1}{2}, & \text{if } x \in [\frac{1}{3}, \frac{2}{3}] \\ \frac{1}{4}, & \text{if } x \in [\frac{1}{9}, \frac{2}{9}] \\ \frac{3}{4}, & \text{if } x \in [\frac{7}{9}, \frac{8}{9}] \\ \vdots & \vdots \end{cases}$$

i.e., on each discarded interval in the Cantor set construction, F is a constant function. Thus F is differentiable with $F'(x) = 0$ for all points of $[0,1] \setminus C$, and because C is a zero set, $F'(x) = 0$ for almost all x.

To show F is continuous, consider $x \in [0, 1]$. If x has base 3 expansion $x = (.x_1 x_2 x_3 \ldots)_3$, then the base 2 expansion of $y = F(x)$ is $y = (.y_1 y_2 y_3 \ldots)_2$, where

$$y_i = \begin{cases} 0, & \text{if there exists } k < i \text{ such that } x_k = 1, \\ 1, & \text{if } x_i = 1 \text{ and there does not exist } k < i \text{ such that } x_k = 1, \\ \frac{x_i}{2}, & \text{if } x_i = 0 \text{ or } x_i = 2 \text{ and there does not exist } k < i \text{ such that } x_k = 1. \end{cases}$$

First notice that $F(x)$ is well defined, because two base 3 expansions of x represent the same number x if and only if x is an endpoint of C. Thus one of its base 3 expansions end in 2's, the other in 0's: $(.x_1 x_2 \ldots x_l 0\bar{2})_3 = x = (.x_1 x_2 \ldots x_l 1\bar{0})_3$. If for some (smallest) $k \leq l$, $x_k = 1$, then $(.y_1 y_2 \ldots)_2 = (.\frac{x_1}{2} \frac{x_2}{2} \ldots \frac{x_{k-1}}{2} 1\bar{0})_2$. If for all $k \leq l$, $x_k \neq 1$, then the two base 2 expansions for $F(x)$ are $(.\frac{x_1}{2} \frac{x_2}{2} \ldots \frac{x_{l-1}}{2} 0\bar{1})_2$ and $(.\frac{x_1}{2} \frac{x_2}{2} \ldots \frac{x_{l-1}}{2} 1\bar{0})_2$. These two base 2 expansions represent the same number y. The same reasoning applies if one has $(.x_1 x_2 \ldots x_l 1\bar{2})_3 = x = (.x_1 x_2 \ldots x_l 2\bar{0})_3$. Thus, $F(x)$ is well defined.

To show $F(x)$ is continuous, let $\varepsilon > 0$ be given. Choose k such that $\frac{1}{2^k} \leq \varepsilon$. If $|x - x_0| < \frac{1}{3^k}$, then there are base 3 expansions of x and x_0 whose first k symbols agree. Therefore, the first k symbols of $F(x)$ and $F(x_0)$ agree as well, which implies that $|F(x) - F(x_0)| \leq \frac{1}{2^{k+1}} < \varepsilon$.

Solution 5.32

In fact, for convenience, we shall prove the result for sequences (f_j) of functions from $[0, 1)$ to $[0, 1)$; this is equivalent to the stated theorem because there is a homomorphism from $\mathbb{R}^+$ to $[0, 1)$.

For a function $f : [0, 1) \to [0, 1)$ we write $D(f)$ for the set of points $x \in [0, 1)$ at which the function f is not feebly continuous, and for a sequence $\mathbf{f} = (f_j)$ of such functions, we write $D(\mathbf{f})$ for the set of points $x \in [0, 1)$ at which the sequence $\mathbf{f}$ is not feebly continuous, so that

$$\bigcup \{D(f_j) : j \in \mathbb{N}\} \subset D(\mathbf{f}).$$

We first *claim* that it suffices to prove that, for each $N \in \mathbb{N}$, the set $D(\mathbf{f})$ is countable for each such sequence $\mathbf{f} = (f_j)$ for which the range, $R(f_j)$, of each function f_j contains at most N points in $[0, 1)$. Indeed, suppose that we have proved that $D(\mathbf{f})$ is countable for each such sequence $\mathbf{f} = (f_j)$, and let $\mathbf{f} = (f_j)$ be any sequence of functions from $[0, 1)$ to $[0, 1)$. For $j, n \in \mathbb{N}$, set

$$f_{j,n}(x) = \frac{1}{n}[n f_j(x)] \quad (x \in [0, 1)),$$

where $[\alpha]$ is the integer part of $\alpha \in \mathbb{R}^+$, so that $|R(f_{n,j})| \leq n$ for each $j \in \mathbb{N}$ and

$$f_{j,n}(y) \leq f_j(y) < f_{j,n}(y) + \frac{1}{n} \quad (y \in [0, 1), j \in \mathbb{N}).$$

Set $\mathbf{f}_n = (f_{n,j} : j \in N)$. By hypothesis, the set $D(\mathbf{f}_n)$ is countable for each $n \in \mathbb{N}$, and so

$$D := \bigcup \{D(\mathbf{f}_n) : n \in \mathbb{N}\}$$

is countable.

Let $x \in [0, 1) \setminus D$, so that each sequence $\mathbf{f}_n$ is feebly continuous at x. We define a sequence (r_n) inductively by requiring that

$$r_n < \min\left\{r_{n-1}, \frac{1}{n}\right\} \quad \text{and} \quad |f_{j,n}(x + r_n) - f_{j,n}(x)| < \frac{1}{n} \quad (j \in \mathbb{N}_n),$$

where $r_0 = 1$, so that $(r_n) \searrow 0$. For each $n \in \mathbb{N}$ and $j \in \mathbb{N}_n$, we have

$$f_j(x + r_n) < f_{j,n}(x + r_n) + \frac{1}{n} < f_{j,n}(x) + \frac{2}{n} < f_j(x) + \frac{2}{n}$$

and, similarly, $f_j(x + r_n) > f_j(x) - 2/n$. For each $j \in \mathbb{N}$ and $n \geq j$, we have

$$|f_j(x + r_n) - f_j(x)| < \frac{2}{n},$$

and so $f_j(x + r_n) \to f_j(x)$ as $n \to \infty$. This establishes the claim.

Now take a sequence $\mathbf{f} = (f_j)$ of functions from $[0, 1)$ to $[0, 1)$ such that $|R(f_j)| \leq k$ for each $j \in \mathbb{N}$, where $k \in \mathbb{N}$.

For each $x \in D(\mathbf{f})$, there exist $j = j(x) \in \mathbb{N}$ and $\delta(x) > 0$ such that
$$\inf\{|f_j(x+t) - f_j(x)| : 0 < t < \delta(x)\} > 0.$$

In particular, $f_j(y) \neq f_j(x)$ whenever $y \in (x, x + \delta(x))$. For each such x, choose $q(x) \in \mathbb{Q} \cap (x, x + \delta(x))$.

Assume toward a contradiction that $D(\mathbf{f})$ is uncountable. Then there exist $j_0 \in \mathbb{N}$ and $q \in \mathbb{Q}$ and such that
$$\left\{x \in [0, 1) : \inf\{|f_{j_0}(x+t) - f_{j_0}(x)| : 0 < t < q\} > 0\right\}$$
is uncountable. In particular, there exist distinct points $x_1, \ldots, x_{k+1} \in [0, 1)$ with $q(x_i) = q$ ($i \in \mathbb{N}_{k+1}$) for which
$$\inf\{|f_{j_0}(x_i+t) - f_{j_0}(x_i)| : 0 < t < q\} > 0;$$
we may suppose that $x_1 < x_2 < \ldots < x_{k+1}$. We note that
$$q \in (x_i, x_i + \delta(x_i)) \quad (i \in \mathbb{N}_{k+1}).$$

Take $i, j \in \mathbb{N}_{k+1}$ with $i < j$. Then $x_i < x_j < q < x_i + \delta(x_i)$, and hence $f_{j_0}(x_j) \neq f_{j_0}(x_i)$. Thus the numbers $f_{j_0}(x_1), \ldots, f_{j_0}(x_{k+1})$ are distinct, a contradiction of the fact that $|R(f_{j_0})| = k$. Thus $D(\mathbf{f})$ is countable, as required.

$\boxed{\text{Solution 5.33}}$

a) Given $c \in \mathbb{R}$ and $\varepsilon > 0$, set $\delta = \dfrac{\varepsilon}{k}$. Then
$$|f(x) - f(y)| < \varepsilon \text{ for all } x \in \mathbb{R} \text{ such that } |x - c| < \delta.$$

Hence if f is a contraction, then f is continuous.

b) To show that (x_n) is Cauchy, observe:
$$|x_{n+1} - x_n| = |f(x_n) - f(x_{n-1})| \leq k|x_n - x_{n-1}| = |f(x_{n-1}) - f(x_{n-2})| \leq k^2|(x_{n-1} - x_{n-2})|$$
we can repeat the above process to obtain
$$|x_{n+1} - x_n| \leq k^n|x_1 - x_0|.$$

Thus if $n > m$,
$$|x_n - x_m| \leq |x_n - x_{n-1}| + |x_{n-1} - x_{n-2}| + \cdots + |x_{m+1} - x_m|$$
$$|x_{n+1} - x_n| \leq (k^{n-1} + k^{n-2} + \cdots + k^m)|x_1 - x_0| \leq \frac{k^m}{1-k}|x_0 - x_1|$$

using the limiting sum of a geometric series, which we may do since $0 < k < 1$. Since $k^m \to 0$ as $m \to \infty$, we must have $|x_n, x_m| < \varepsilon$ for any $\varepsilon > 0$ whenever m and n are sufficiently large. Hence (x_n) is a Cauchy sequence. Since $\mathbb{R}$ is a complete space, every Cauchy sequence converges in it, and the existence of $\lim_{n \to \infty} x_n$ is assured. We set $\lim_{n \to \infty} x_n = x^*$.

c) To show x^* is a fixed point of f, we note that, for any positive integer n,

$$0 \leq |f(x^*)-x^*| \leq |f(x^*)-x_n|+|x_n-x^*| = |f(x^*)-f(x_{n-1})|+|x_n-x^*| \leq k|x^*-x_{n-1}|+|x_n-x^*|$$

and so $|f(x^*) - x^*| = 0$ since $|x_n - x^*| \to 0$ (and $|x^* - x_{n-1}| \to 0$). Thus $f(x^*) = x^*$.

d) Suppose there are two fixed points say x^* and $\tilde{x}$, so also $f(\tilde{x}) = \tilde{x}$. Then

$$|x^* - \tilde{x}| = |f(x^*) - f(\tilde{x})| \leq k|x^* - \tilde{x}|$$

which since $k < 1$ can only be true when $|x^* - \tilde{x}| = 0$; that is, $x^* = \tilde{x}$. Hence there is a unique fixed point of f.

Chapter 6

Differentiability

[His epitaph:] Who, by vigor of mind almost divine, the motions and figures of the planets, the paths of comets, and the tides of the seas first demonstrated.

Isaac Newton (1643–1727)

- Let f be a real-valued function defined on an interval I containing the point c. We say f is *differentiable* at c if the limit

$$\lim_{x \to c} \frac{f(x) - f(c)}{x - c} = \lim_{h \to 0} \frac{f(c+h) - f(c)}{h}$$

exists. In this case we write $f'(c)$ for this limit. If the function f is differentiable at each point of the set $S \subseteq I$, then f is said to be differentiable on S and the function $f' : S \to R$ is called the derivative of f on S. When f is differentiable at c the *tangent line* to f at c is the linear function $L(x) = f(c) + f'(c)(x - c)$.

- *Rolle's Theorem*: Let f be a continuous function on $[a, b]$ that is differentiable on (a, b) and such that $f(a) = f(b) = 0$. Then there exists at least one point $c \in (a, b)$ such that $f'(c) = 0$.

- *Mean Value Theorem*: If $f : [a, b] \to \mathbb{R}$ is continuous on $[a, b]$ and differentiable on (a, b), then there exists a point $c \in (a, b)$ where

$$f'(c) = \frac{f(b) - f(a)}{b - a}.$$

- If f is differentiable on (a,b) and f' is continuous, we say f *is of class C^1*. If f' is differentiable, then we can get the second derivative f''. If f'' is continuous, we say f is *of class C^2*, and so on.

- *Taylor's Theorem*: Let $f : [a,b] \to \mathbb{R}$ be of class C^n (n is a positive integer) and let $x_0 \in [a,b]$. Then for each $x \in [a,b]$ with $x \neq x_0$ there exists a point c between x and x_0 such that

$$f(x) = f(x_0) + f'(x_0)(x-x_0) + \frac{f''(x_0)}{2!}(x-x_0)^2 + \cdots + \frac{f^{(n)}(x_0)}{n!}(x-x_0)^n + \frac{f^{(n+1)}(c)}{(n+1)!}(x-x_0)^{n+1}$$

Problem 6.1 Show that $f(x) = |x|$ is not differentiable at 0.

Problem 6.2 Discuss the differentiability of $f(x) = |x^2 - 4|$ at $x = 2$.

Problem 6.3 Prove that every differentiable function is continuous.

Problem 6.4 Prove that if the function $f : I \to \mathbb{R}$ has a bounded derivative on I, then f is uniformly continuous on I. Is the converse true?

Problem 6.5 Let f be a function on $[a,b]$ that is differentiable at c. Let $L(x)$ be the tangent line to f at c. Prove that l is the unique linear function with the property that

$$\lim_{x \to c} \frac{f(x) - L(x)}{x - c} = 0.$$

Problem 6.6 Determine whether or not f is differentiable at 0:

1. $f(x) = \sqrt[3]{x}$
2. $f(x) = \sqrt{|x|}$
3. $f : \mathbb{R} \to \mathbb{R}$ defined by $f(x) = x \sin\left(\frac{1}{x}\right)$ if $x \neq 0$ and $f(0) = 0$
4. $f(x) = x|x|$

CHAPTER 6. DIFFERENTIABILITY

Problem 6.7 Discuss the differentiability of

$$f(x) = \begin{cases} x^2 \sin\left(\dfrac{1}{x}\right) & \text{if } x \neq 0, \\ 0 & \text{if } x = 0 \end{cases}$$

at $x = 0$.

Problem 6.8 Find the derivatives (if they exist) of

$$f(x) = \begin{cases} x^2 e^{-x^2} & \text{if } |x| \leq 1, \\ \dfrac{1}{e} & \text{if } |x| > 1. \end{cases}$$

Problem 6.9 Suppose a differentiable function $f : \mathbb{R} \longrightarrow \mathbb{R}$ has a uniformly continuous derivative on $\mathbb{R}$. Show that

$$\lim_{n \to \infty} n\left[f\left(x + \frac{1}{n}\right) - f(x) \right] = f'(x).$$

Problem 6.10 Let $f(x)$ be differentiable at a. Find

$$\lim_{n \to \infty} \frac{a^n f(x) - x^n f(a)}{x - a}$$

where $n \in \mathbb{N}$.

Problem 6.11 We say a function $f : (a, b) \to \mathbb{R}$ is uniformly differentiable if f is differentiable on (a, b) and for each $\varepsilon > 0$ there exists a $\delta > 0$ such that

$$0 < |x - y| < \delta \quad \text{and} \quad x, y \in (a, b) \quad \Rightarrow \quad \left| \frac{f(x) - f(y)}{x - y} - f'(x) \right| < \varepsilon.$$

Prove that if f is uniformly differentiable, then f' is continuous. Then give an example of a function that is differentiable but not uniformly differentiable.

Problem 6.12 Let $f : (-a, a) \to \mathbb{R}$, with $a > 0$. Assume $f(x)$ is continuous at 0 and such that the limit

$$\lim_{x \to 0} \frac{f(x) - f(\kappa\, x)}{x} = l$$

exists, where $0 < \kappa < 1$. Show that $f'(0)$ exists. What happens to this conclusion when $\kappa > 1$?

Problem 6.13 Let $f: \mathbb{R} \to \mathbb{R}$ be a continuous function. Consider the sequence
$$x_0 \in \mathbb{R} \text{ and } x_{n+1} = f(x_n).$$
Assume that $\lim_{n \to \infty} x_n = l$ and $f'(l)$ exists. Show that $|f'(l)| \leq 1$.

Problem 6.14 Let $f(x) = \sqrt{1+x^2}$. Show that
$$(1+x^2)f^{[n+2]}(x) + (2n+1)xf^{[n+1]}(x) + (n^2-1)f^{[n]}(x) = 0,$$
for any $n \geq 1$. Use this identity to show that $f^{[2n+1]}(0) = 0$ for any $n \geq 0$. The notation $f^{[m]}(x)$ denotes the mth derivative of f.

Problem 6.15 Show that the equation $e^x = 1-x$ has one solution in $\mathbb{R}$. Find this solution.

Problem 6.16 Let $f: [a,b] \to \mathbb{R}$ be continuous and differentiable everywhere in (a,b) except maybe at $c \in (a,b)$. Assume that
$$\lim_{x \to c} f'(x) = l.$$
Show that $f(x)$ is differentiable at c and $f'(c) = l$.

Problem 6.17 Let $f: \mathbb{R} \to \mathbb{R}$ differentiable everywhere. Let $x_0 \in \mathbb{R}$ and $h \in \mathbb{R}$. Show that there exists $\theta \in (0,1)$ such that
$$f(x_0+h) - f(x_0) = hf'(x_0 + \theta h).$$
Set $f(x) = \dfrac{1}{1+x}$ and $x_0 > 0$. Find the limit of θ when $h \to 0$.

Problem 6.18 Show the following inequalities:

(a) $\ln(1+x) \leq x$, for any $x \geq 0$;

(b) $x + \dfrac{x^3}{3} \leq \tan(x)$, for any $x \in \left(0, \dfrac{\pi}{2}\right)$;

(c) $x - \dfrac{x^3}{6} < \sin(x) < x$, for $0 < x \leq \dfrac{\pi}{2}$;

(d) $1 - \dfrac{x^2}{2} < \cos(x) < 1 - \dfrac{x^2}{2} + \dfrac{x^4}{24}$, for $0 < x \leq \dfrac{\pi}{2}$.

CHAPTER 6. DIFFERENTIABILITY

Problem 6.19 Let $f(x)$ be a continuous function on $[a,b]$, differentiable on (a,b), and $f'(x) \neq 0$ for any $x \in (a,b)$. Show that $f(x)$ is one-to-one. Then show that $f'(x) > 0$ for every $x \in (a,b)$, or $f'(x) < 0$ for every $x \in (a,b)$. Deduce from this that $f'(x)$ satisfies the Intermediate Value Theorem without use of any continuity of $f'(x)$.

Problem 6.20 Let $f : \mathbb{R} \to \mathbb{R}$ be a differentiable function. Suppose that $f'(x) > f(x)$ for all $x \in \mathbb{R}$, and $f(x_0) = 0$ for some $x_0 \in \mathbb{R}$. Prove that $f(x) > 0$ for all $x > x_0$. As an application of this, show that
$$ae^x = 1 + x + \frac{x^2}{2}$$
where $a > 0$ has exactly one root.

Problem 6.21 Let $f : \mathbb{R} \to \mathbb{R}$. Assume that for any $x, t \in \mathbb{R}$ we have
$$|f(x) - f(t)| \leq |x - t|^{1+\alpha}$$
where $\alpha > 0$. Show that $f(x)$ is constant.

Problem 6.22 Let $f : [0, \infty) \to \mathbb{R}$ differentiable everywhere. Assume that $\lim_{x \to \infty} f(x) + f'(x) = 0$. Show that $\lim_{x \to \infty} f(x) = 0$.

Problem 6.23 Let $f : [0,1] \to \mathbb{R}$ be continuous and differentiable inside $(0,1)$ such that

(i) $f(0) = 0$,

(ii) and there exists $M > 0$ such that $|f'(x)| \leq M|f(x)|$, for $x \in (0,1)$.

Show that $f(x) = 0$ for $x \in [0,1]$.

Problem 6.24 Consider the function
$$f(x) = \begin{cases} e^{-1/x^2} & \text{if } x \neq 0, \\ 0 & \text{if } x = 0. \end{cases}$$
Show that $f^{[n]}(0) = 0$, for $n = 1, 2, \ldots$.

Problem 6.25 Consider a function $f(x)$ whose second derivative $f''(x)$ exists and is continuous on (a,b). Let $c \in (a,b)$. Show that

$$\lim_{h \to 0} \frac{f(c+h) - 2f(c) + f(c-h)}{h^2} = f''(c).$$

Is the existence of the second derivative necessary to prove the existence of the above limit?

Problem 6.26 Let f be a real-valued twice continuously differentiable function on $[a,b]$. Let $\widetilde{x}$ be a simple zero of f in (a,b). Show that Newton's method defined by

$$x_{n+1} = g(x_n) \text{ and } g(x_n) = x_n - \frac{f(x_n)}{f'(x_n)}$$

is a contraction in some neighborhood of $\widetilde{x}$, so that the iterative sequence converges to $\widetilde{x}$ for any x_0 sufficiently close to $\widetilde{x}$.

Note that $\widetilde{x}$ is a simple zero implies that $f'(x) \neq 0$ on some neighborhood U of $\widetilde{x}$ where $U \subset [a,b]$ and f'' is bounded on U.

Problem 6.27 Consider a function $f(x)$ whose second derivative $f''(x)$ exists and is continuous on $[0,1]$. Assume that $f(0) = f(1) = 0$ and suppose that there exists $A > 0$ such that $|f''(x)| \leq A$ for $x \in [0,1]$. Show that

$$\left| f'\left(\frac{1}{2}\right) \right| \leq \frac{A}{4} \text{ and } |f'(x)| \leq \frac{A}{2}$$

for $0 < x < 1$.

Problem 6.28 Let I be an open interval, and let $f : I \to \mathbb{R}$ be such that the nth order derivative $f^{(n)}$ exists on I. Show that for $a \in I$ and for each h such that $a + h \in I$, we have

$$f(a+h) = f(a) + hf'(a) + \frac{h^2}{2!}f''(a) + \cdots + \frac{h^{n-1}}{(n-1)!}f^{(n-1)}(a) + R_n,$$

where $R_n = \dfrac{h^n}{n!} f^{(n)}(a + \theta h)$ for some $\theta \in (0,1)$ (Lagrange's form of the remainder)

and $R_n = \dfrac{h^n (1-\theta)^{n-1}}{(n-1)!} f^{(n)}(a + \theta h)$ for some $\theta \in (0,1)$ (Cauchy's form of the remainder).

Note that here $f'', f''', \ldots, f^{(j)}$ are the 2nd, 3rd, ..., jth derivatives of f, θ depends on h, and is in principle different in Lagrange's and Cauchy's forms: $a + \theta h$ for $\theta \in (0,1)$ is a convenient way of specifying some number c between a and $a + h$ (even when h is negative). If $n = 1$, the two forms are equal, and agree with $hf'(c)$ in the Mean Value Theorem.

CHAPTER 6. DIFFERENTIABILITY

Problem 6.29 (**General Binomial Theorem**) Show that for fixed $s \in \mathbb{R}$, the power series expansion

$$(1+x)^s = 1 + sx + \frac{s(s-1)}{2!}x^2 + \frac{s(s-1)(s-2)}{3!}x^3 + \cdots + \frac{s(s-1)\cdots(s-k+1)}{k!}x^k + \cdots$$

is valid

(i) for all x whenever $s \in \mathbb{N}$,

(ii) for $|x| < 1$ in all cases,

(iii) for $x = 1$ if and only if $s > -1$,

(iv) for $x = -1$ if and only if $s \geq 0$.

More generally, if $a > 0$, then the expansion $(a+x)^s = a^s + sxa^{s-1} + \frac{s(s-1)}{2!}x^2 a^{s-2} + \cdots$ is valid whenever $|x| < a$.

Hint: Use the Maclaurin series of $f(x) = (1+x)^s$ together with Cauchy's form of the remainder R_n.

Problem 6.30 (**The Leibnitz Formula**) Let $f(x)$ and $g(x)$ be two differentiable functions with continuous nth derivatives. Then their product has a continuous nth derivative, and

$$\frac{d^n}{dx^n}\Big(f(x)g(x)\Big) = \sum_{k=0}^{n} \binom{n}{k} \frac{d^k}{dx^k}\Big(f(x)\Big) \frac{d^{n-k}}{dx^{n-k}}\Big(g(x)\Big).$$

Problem 6.31 (**Implicit Function Theorem**) Let $D = \{(x,y) \in \mathbb{R}^2 : a \leq x \leq b\}$ and $F : D \to \mathbb{R}$ be a function where its partial derivative with respect to y exists and there exist $m, M > 0$ such that

$$0 < m < \frac{\partial F}{\partial y} \leq M \text{ for all } (x,y) \in D.$$

Show that there exists one and only one continuous function $y(x)$ on $[a,b]$ such that

$$F(x, y(x)) = 0.$$

Note: This means the equation $F(x, y(x)) = 0$ does implicitly define a unique continuous function y in terms of x. To solve this problem consider the vector space $C[a,b]$ of all continuous real-valued functions defined on $[a,b]$ with $\| f \| = \max_{a \leq x \leq b} |f(x)|$ and define a map

$$T : (C[a,b], \| \cdot \|) \to (C[a,b], \| \cdot \|)$$

as

$$Ty(x) = y(x) - \frac{1}{M}F(x, y(x)).$$

Show T is a contraction and use the Banach Contraction Mapping Theorem.

Problem 6.32 (Legendre polynomials)
Consider the sequence of polynomial functions

$$L_n = \frac{1}{2^n n!} \frac{d^n}{dx^n}(x^2-1)^n, \ n \in \mathbb{N}.$$

1. Show the recurrence relationship

$$(n+1)L_{n+1} = (2n+1)xL_n - nL_{n-1},$$

   for any $n \geq 1$.

2. Find the degree of L_n, $n \in \mathbb{N}$.

3. Show that for any polynomial P, with degree less than or equal to $n-1$, we have

$$\int_{-1}^{1} L_n(x)P(x)dx = 0.$$

4. Find $\int_{-1}^{1} \left(L_n(x)\right)^2 dx$.

Solutions

Solution 6.1

Let $x_n = \frac{(-1)^n}{n}$ for $n \in \mathbb{N}$. Then the sequence (x_n) converges to 0, but the corresponding sequence of quotients does not converge.

If n is even $x_n = \frac{1}{n}$, so that

$$\frac{f(x_n) - f(0)}{x_n - 0} = \frac{\frac{1}{n} - 0}{\frac{1}{n} - 0} = 1.$$

If n is odd, we have $x_n = -\frac{1}{n}$, so that

$$\frac{f(x_n) - f(0)}{x_n - 0} = \frac{\frac{1}{n} - 0}{-\frac{1}{n} - 0} = -1.$$

Since the two subsequences have different limits, the sequence $\frac{f(x_n)-f(0)}{x_n-0}$ does not converge. Thus f is not differentiable at zero.

Solution 6.2

By definition of the derivative, we know that $f'(2)$ exists if and only if $\lim\limits_{x \to 2} \frac{f(x) - f(2)}{x - 2}$ exists. Since

$$\lim_{x \to 2} \frac{f(x) - f(2)}{x - 2} = \lim_{x \to 2} \frac{|x^2 - 4|}{x - 2},$$

we will consider the side limits in order to take care of the absolute value. So

$$\lim_{x \to 2+} \frac{|x^2 - 4|}{x - 2} = \lim_{x \to 2+} \frac{(x^2 - 4)}{x - 2} = \lim_{x \to 2+} x + 2 = 4.$$

On the other hand, we have

$$\lim_{x \to 2-} \frac{|x^2 - 4|}{x - 2} = \lim_{x \to 2-} \frac{-(x^2 - 4)}{x - 2} = \lim_{x \to 2+} -(x + 2) = -4.$$

So $\lim\limits_{x \to 2} \frac{f(x) - f(2)}{x - 2}$ does not exist, i.e., $f(x)$ is not differentiable at $x = 2$. Note that the graph of $f(x)$ has two half-tangents at $x = 2$ with different slopes.

Solution 6.3

Suppose that f is differentiable at $x = c$; we compute

$$\begin{aligned}\lim_{x \to c} f(x) &= \lim_{x \to c} f(x) + (x - c)\frac{f(x) - f(c)}{x - c} \\ &= f(c) + 0 \cdot f'(c) = f(c).\end{aligned}$$

Solution 6.4

By hypothesis, we know that there is a constant $M > 0$ such that $|f'(x)| \leq M$ for all $x \in I$. Using the Mean Value Theorem, we have

$$f(x) - f(y) = f'(a)(x - y),$$

where a is between x and y for $x, y \in I$. Hence, $|f(x) - f(y)| \leq M|x - y|$. Given $\varepsilon > 0$, choose $\delta = \varepsilon/M$. Then $|x - y| < \delta$ implies $|f(x) - f(y)| \leq M|x - y| < M\delta = \frac{M\varepsilon}{M} = \varepsilon$. Hence, f is uniformly continuous on the interval I. The converse is not true. Indeed, consider the function $f(x)$ defined on $[0, 1]$ by

$$f(x) = \begin{cases} x^2 \sin\left(\dfrac{1}{x^2}\right) & \text{if } 0 < x \leq 1, \\ 0 & \text{if } x = 0. \end{cases}$$

Since $|x^2 \sin(1/x^2)| \leq |x^2|$, we see that f is continuous at 0. Since f is continuous on $(0, 1]$, we conclude that f is continuous on $[0, 1]$. Then $f(x)$ is uniformly continuous on $[0, 1]$. However, the derivative

$$f'(x) = 2x \sin \frac{1}{x^2} - \frac{2}{x} \cos \frac{1}{x^2}$$

is unbounded on $(0, 1)$, because if $x_n = \frac{1}{\sqrt{2\pi n}}$, then $x_n \to 0$ and $|f'(x_n)| = 2\sqrt{2\pi n} \to \infty$ as $n \to \infty$.

Solution 6.5

Setting $h = x - c$,

$$\begin{aligned} \lim_{x \to c} \frac{f(x) - L(x)}{x - c} &= \lim_{h \to 0} \frac{f(c+h) - (f(c) + f'(c)h)}{h} \\ &= \lim_{h \to 0} \frac{f(c+h) - f(c)}{h} - f'(c) \\ &= 0. \end{aligned}$$

Now if we have another function $K(x)$ that satisfies this limit, then by the continuity of K and f we have

$$\begin{aligned} f(c) - K(c) &= \lim_{x \to c} f(x) - K(c) \\ &= \lim_{x \to c} (x - c) \frac{f(x) - K(c)}{(x - c)} \\ &= 0. \end{aligned}$$

Thus $K(x) = f(c) + m(x - c)$ where m is the slope and

$$\begin{aligned} m = K'(c) &= \lim_{h \to 0} \frac{K(c+h) - K(c)}{h} \\ &= \lim_{h \to 0} \frac{K(c+h) - f(c+h)}{h} + \frac{f(c+h) - f(c)}{h} \\ &= 0 + f'(c) = f'(c). \end{aligned}$$

Thus the line K that goes through the point $(c, f(c))$ has the same slope as L. Therefore $K = L$.

Solution 6.6

1. $f'(x) = \frac{1}{3}x^{-2/3}$ for $x \neq 0$. But at $x = 0$ we have
$$\lim_{h \to 0} \frac{f(h) - f(0)}{h} = \lim_{h \to 0} h^{-2/3} = +\infty,$$
which is not differentiable at the origin. (Vertical tangent at the origin.)

2. $f(x) = \sqrt{|x|}$ has a cusp, with right derivative of $+\infty$ and left derivative of $-\infty$. f is not differentiable at the origin; that is,
$$f'(x) = \begin{cases} \frac{1}{2\sqrt{x}} & \text{if } x > 0, \\ -\frac{1}{2\sqrt{-x}} & \text{if } x < 0, \end{cases}$$
since
$$f'_+(0) = \lim_{h \to 0^+} \frac{\sqrt{h} - 0}{h} = +\infty \quad \text{and} \quad f'_-(0) = \lim_{h \to 0^-} \frac{\sqrt{-h} - 0}{h} = -\infty.$$

3. For $x \neq 0$, we have
$$\frac{f(x) - f(0)}{x - 0} = \frac{x \sin \frac{1}{x}}{x} = \sin \frac{1}{x}.$$
But $\lim_{x \to 0} \sin\left(\frac{1}{x}\right)$ does not exist. Thus f is not differentiable at $x = 0$.

4. Observe that
$$f(x) = \begin{cases} x^2 & \text{if } x \geq 0, \\ -x^2 & \text{if } x < 0, \end{cases}$$
therefore
$$f'(x) = \begin{cases} 2x & \text{if } x > 0, \\ -2x & \text{if } x < 0, \end{cases}$$
and
$$f'_+(0) = \lim_{h \to 0^+} \frac{h^2 - 0}{h} = 0 = -f'_-(0).$$
f is differentiable at 0.

Solution 6.7

Since the sine function is bounded and $\lim_{x \to 0} x^2 = 0$, we have
$$\lim_{x \to 0} x^2 \sin\left(\frac{1}{x}\right) = 0 = f(0).$$

Hence $f(x)$ is continuous at $x = 0$. In order to find out if $f(x)$ is differentiable at $x = 0$, let us investigate the limit $\lim_{x \to 0} \frac{f(x) - f(0)}{x - 0}$. Since
$$\lim_{x \to 0} \frac{f(x) - f(0)}{x - 0} = \lim_{x \to 0} \frac{f(x)}{x} = \lim_{x \to 0} x \sin\left(\frac{1}{x}\right) = 0.$$

Therefore $f(x)$ is differentiable at $x = 0$ and $f'(0) = 0$. Note that if one wants to use the properties of the derivatives, we will get

$$f'(x) = 2x \sin\left(\frac{1}{x}\right) - \cos\left(\frac{1}{x}\right),$$

which does not have a limit when $x \to 0$.

Solution 6.8

Clearly
$$f'(x) = \begin{cases} 2xe^{-x^2}(1-x^2) & \text{if } |x| < 1, \\ 0 & \text{if } |x| > 1, \end{cases}$$

for x general. In addition, at $x = 1$ we have that

$$f'_+(1) = \lim_{x \to 1^+} \frac{\frac{1}{e} - \frac{1}{e}}{x-1} = 0 \quad \text{and} \quad f'_-(1) = \lim_{x \to 1^-} \frac{x^2 e^{-x^2} - \frac{1}{e}}{x-1} = \frac{d}{dx}(x^2 e^{-x^2})\bigg|_{x=1} = 0.$$

Solution 6.9

Since f' is uniformly continuous on $\mathbb{R}$, given $\varepsilon > 0$ there exists $\delta > 0$ such that $|f'(x) - f'(y)|\varepsilon$ for any $x, y \in \mathbb{R}$ for which $|x - y| < \delta$. Let $N \in \mathbb{N}$ such that for any $n \geq N$ we have $1/n < \delta$. Then for any $x \in \mathbb{R}$ we have

$$|f'(t) - f'(x)| < \varepsilon, \quad \text{for any } t \in \left(x, x + \frac{1}{n}\right).$$

Since f is differentiable, we can use the Mean Value Theorem to obtain

$$\left| n\left[f\left(x + \frac{1}{n}\right) - f(x)\right] - f'(x) \right| = \left| \frac{f\left(x + \frac{1}{n}\right) - f(x)}{\frac{1}{n}} - f'(x) \right| = |f'(t_n) - f'(x)| < \varepsilon$$

for some $t_n \in \left(x, x + \frac{1}{n}\right)$, which yields

$$\lim_{n \to \infty} n\left[f\left(x + \frac{1}{n}\right) - f(x)\right] = f'(x).$$

Solution 6.10

We have

$$\frac{a^n f(x) - x^n f(a)}{x - a} = \frac{a^n f(x) - a^n f(a) + a^n f(a) - x^n f(a)}{x - a} = a^n \frac{f(x) - f(a)}{x - a} - f(a) \frac{x^n - a^n}{x - a}.$$

Since $f(x)$ and x^n are both differentiable at a, we get

$$\lim_{n \to \infty} \frac{a^n f(x) - x^n f(a)}{x - a} = a^n f'(a) - f(a) n a^{n-1}.$$

CHAPTER 6. DIFFERENTIABILITY

Solution 6.11

Suppose f is uniformly differentiable. Let $\varepsilon > 0$ be given. Then there exists some $\delta > 0$ such that

$$0 < |x - y| < \delta \quad \Rightarrow \quad \left| \frac{f(x) - f(y)}{x - y} - f'(x) \right| < \frac{\varepsilon}{2}.$$

Therefore,

$$|f'(x) - f'(y)| = \left| f'(x) - \frac{f(x) - f(y)}{x - y} \right| + \left| \frac{f(y) - f(x)}{y - x} - f'(y) \right| \leq \frac{\varepsilon}{2} + \frac{\varepsilon}{2} = \varepsilon.$$

Hence, f' is continuous. Now consider

$$f(x) = \begin{cases} x^2 \sin \frac{1}{x} & \text{if } x \neq 0, \\ 0 & \text{if } x = 0. \end{cases}$$

This function is differentiable everywhere but $f'(x) = 2x \sin \frac{1}{x} - \cos \frac{1}{x}$ for $x \neq 0$ and $\lim_{x \to 0} f'(x)$ does not exist. Thus, f' exists but is not continuous.

Solution 6.12

Let $\varepsilon > 0$. Set $\varepsilon^* = \varepsilon(1 - \kappa)$. Then we have $\varepsilon^* > 0$. Since $\lim_{x \to 0} \dfrac{f(x) - f(\kappa x)}{x} = l$, there exists $\delta > 0$ such that for $x \in (-a, a)$ and $|x| < \delta$, we have

$$l - \varepsilon^* < \frac{f(x) - f(\kappa x)}{x} < l + \varepsilon^*.$$

Since $\kappa \in (0, 1)$, we have $|\kappa^n x| \leq |x| < \delta$ for any $x \in (-\delta, \delta) \cap (-a, a)$ and $n \geq 1$. So let $x \in (-\delta, \delta) \cap (-a, a)$ be fixed. Hence for any $n \geq 0$, we have

$$l - \varepsilon^* < \frac{f(\kappa^n x) - f(\kappa^{n+1} x)}{\kappa^n x} < l + \varepsilon^*,$$

or

$$(l - \varepsilon^*)\kappa^n < \frac{f(\kappa^n x) - f(\kappa^{n+1} x)}{x} < (l + \varepsilon^*)\kappa^n.$$

So

$$\sum_{n=0}^{N} (l - \varepsilon^*)\kappa^n < \sum_{n=0}^{N} \frac{f(\kappa^n x) - f(\kappa^{n+1} x)}{x} < \sum_{n=0}^{N} (l + \varepsilon^*)\kappa^n,$$

for any $N \geq 1$. Since

$$\begin{cases} \displaystyle\sum_{n=0}^{N} (l - \varepsilon^*)\kappa^n &= (l - \varepsilon^*)\dfrac{1 - \kappa^{N+1}}{1 - \kappa}, \\[2ex] \displaystyle\sum_{n=0}^{N} (l + \varepsilon^*)\kappa^n &= (l + \varepsilon^*)\dfrac{1 - \kappa^{N+1}}{1 - \kappa}, \\[2ex] \displaystyle\sum_{n=0}^{N} \dfrac{f(\kappa^n x) - f(\kappa^{n+1} x)}{x} &= \dfrac{f(x) - f(\kappa^{N+1} x)}{x}, \end{cases}$$

then
$$(l-\varepsilon^*)\frac{1-\kappa^{N+1}}{1-\kappa} < \frac{f(x)-f(\kappa^{N+1}\,x)}{x} < (l+\varepsilon^*)\frac{1-\kappa^{N+1}}{1-\kappa},$$

for any $N \geq 1$. Since $\lim_{N\to\infty} \kappa^N = 0$, we get by letting $N \to \infty$,

$$(l-\varepsilon^*)\frac{1}{1-\kappa} \leq \frac{f(x)-f(0)}{x} \leq (l+\varepsilon^*)\frac{1}{1-\kappa}.$$

Using the definition of ε^*, we get

$$\frac{l}{1-\kappa} - \varepsilon \leq \frac{f(x)-f(0)}{x} \leq \frac{l}{1-\kappa} + \varepsilon,$$

for any $x \in (-\delta, \delta) \cap (-a, a)$. This obviously implies

$$\lim_{x\to 0} \frac{f(x)-f(0)}{x} = \frac{l}{1-\kappa}.$$

Therefore $f(x)$ is differentiable at 0. If we assume $\kappa > 1$, then $0 < \frac{1}{\kappa} < 1$. Easy algebraic manipulations will imply

$$\lim_{x\to 0} \frac{f(x)-f(\kappa\,x)}{x} = \lim_{t\to 0} \frac{f\left(\frac{t}{\kappa}\right) - f(t)}{\frac{t}{\kappa}},$$

where $t = \kappa\,x$. Hence

$$\lim_{x\to 0} \frac{f(x)-f(\kappa\,x)}{x} = \lim_{t\to 0} \kappa \frac{f\left(\frac{t}{\kappa}\right) - f(t)}{t} = -\kappa \lim_{t\to 0} \frac{f(t) - f\left(\frac{t}{\kappa}\right)}{t}.$$

So

$$\lim_{x\to 0} \frac{f(x) - f\left(\frac{x}{\kappa}\right)}{x} = -\frac{l}{\kappa}.$$

The proof above will then imply that $f(x)$ is differentiable at 0 and that

$$\lim_{x\to 0} \frac{f(x)-f(0)}{x} = \frac{-\frac{l}{\kappa}}{1-\frac{1}{\kappa}} = \frac{l}{1-\kappa}.$$

Therefore the conclusion in the statement is independent from the conditions $\kappa < 1$ or $\kappa > 1$.

Solution 6.13

Assume not, i.e., $|f'(l)| > 1$. First note that since $f(x)$ is continuous and $x_{n+1} = f(x_n)$, for any $n \geq 1$, we get by letting $n \to \infty$, $f(l) = l$. Since

$$\lim_{x\to l} \frac{f(x)-f(l)}{x-l} = f'(l),$$

CHAPTER 6. DIFFERENTIABILITY

then
$$\lim_{x \to l} \left| \frac{f(x) - l}{x - l} \right| = |f'(l)| > 1 \,.$$

Take $\varepsilon = \dfrac{|f'(l)| - 1}{2} > 0$. Then there exists $\delta > 0$ such that for any $x \in (l - \delta, l + \delta)$, we have

$$|f'(l)| - \varepsilon < \left| \frac{f(x) - l}{x - l} \right| < |f'(l)| + \varepsilon,$$

or
$$\frac{|f'(l)| + 1}{2} < \left| \frac{f(x) - l}{x - l} \right| \,.$$

In particular, since $1 < |f'(l)|$, we have

$$|x - l| < |x - l| \frac{|f'(l)| + 1}{2} < |f(x) - l|$$

for any $x \in (l - \delta, l + \delta)$. Since $\{x_n\}$ converges to l, there exists $N \geq 1$ such that for any $n \geq N$, we have $x_n \in (l - \delta, l + \delta)$. So

$$|x_n - l| < |f(x_n) - l| = |x_{n+1} - l|$$

for any $n \geq N$. In particular, we have

$$|x_N - l| < |x_{N+1} - l| < |x_n - l|$$

for any $n > N$. If we let $n \to \infty$, we will get

$$|x_N - l| < |x_{N+1} - l| \leq 0$$

which generates the desired contradiction. So we must have $|f'(l)| \leq 1$. Note that one may think that maybe $|f'(l)| < 1$. That is not the case in general. Indeed, take $f(x) = \sin(x)$. Then $\{x_n\}$ converges to 0 but $f'(0) = \cos(0) = 1$.

Solution 6.14

Note first that
$$f'(x) = \frac{2x}{2\sqrt{1 + x^2}} = \frac{x}{\sqrt{1 + x^2}} \,.$$

Hence $(1 + x^2) f'(x) = x\sqrt{1 + x^2} = x f(x)$. If we take the derivative of both sides of this equation, we get

$$2x f'(x) + (1 + x^2) f''(x) = f(x) + x f'(x)$$

or
$$(1 + x^2) f''(x) + x f'(x) - f(x) = 0 \,.$$

Again let us take the derivative of both sides to get

$$(1 + x^2) f^{[3]}(x) + 2x f''(x) + x f''(x) + f'(x) - f'(x) = 0$$

or
$$(1 + x^2) f^{[3]}(x) + 3x f''(x) = 0 \,.$$

So the desired identity is valid when $n = 1$. Assume it is still valid for n and let us prove it for $n + 1$. So we have

$$(1 + x^2)f^{[n+2]}(x) + (2n + 1)xf^{[n+1]}(x) + (n^2 - 1)f^{[n]}(x) = 0.$$

If we take the derivative of both sides of this equation, we get

$$(1 + x^2)f^{[n+3]}(x) + 2xf^{[n+2]}(x) + (2n + 1)xf^{[n+2]}(x) + (2n + 1)f^{[n+1]}(x) + (n^2 - 1)f^{[n+1]}(x) = 0,$$

or

$$(1 + x^2)f^{[n+3]}(x) + (2n + 3)xf^{[n+2]}(x) + \left(2n + 1 + n^2 - 1\right)f^{[n+1]}(x) = 0.$$

Since $(n + 1)^2 - 1 = 2n + 1 + n^2 - 1$, we get

$$(1 + x^2)f^{[n+3]}(x) + (2n + 3)xf^{[n+2]}(x) + \left((n + 1)^2 - 1\right)f^{[n+1]}(x) = 0.$$

Hence by induction the desired identity is true for any $n \geq 1$. First note that $f(0) = 1$, $f'(0) = 1$. From the identity

$$(1 + x^2)f''(x) + xf'(x) - f(x) = 0$$

we get $f''(0) = 1$. Assume that $f^{[2n+1]}(0) = 0$. Then from the identity

$$(1 + x^2)f^{[2n+3]}(x) + (2n + 1)xf^{[2n+2]}(x) + \left((2n + 1)^2 - 1\right)f^{[2n+1]}(x) = 0,$$

we get

$$f^{[2n+3]}(0) + (2n + 1)0 f^{[2n+2]}(0) + \left((2n + 1)^2 - 1\right)f^{[2n+1]}(0) = 0,$$

or $f^{[2n+3]}(0) + \left((2n + 1)^2 - 1\right)f^{[2n+1]}(0) = 0$. Since $f^{[2n+1]}(0) = 0$, we get $f^{[2n+3]}(0) = 0$. By induction we deduce that $f^{[2n+1]}(0) = 0$, for any $n \geq 0$.

$\boxed{\text{Solution 6.15}}$

Since $e^0 = 1 - 0 = 1$, then 0 is one solution of the equation $e^x = 1 - x$. Let us prove that this equation has only one solution. Assume not. Then there exist at least two solutions $x_1 < x_2$. They must satisfy

$$f(x_1) = f(x_2) = 0$$

where $f(x) = e^x + x - 1$. Since this function is differentiable everywhere, we can use Rolle's theorem, there exists $c \in (x_1, x_2)$ such that $f'(c) = 0$. But $f'(x) = e^x + 1$ so $e^c = -1$ which is not true. Therefore our original equation has one solution, i.e., $x = 0$.

$\boxed{\text{Solution 6.16}}$

Without loss of generality take $x \in (c, b)$. Then by the Mean Value Theorem, there exists $\theta_x \in (c, x)$ such that

$$\frac{f(x) - f(c)}{x - c} = f'(\theta_x).$$

When $x \to c$, the Squeeze Theorem implies $\theta_x \to c$ as well. So

$$\lim_{x \to c+} f'(\theta_x) = l,$$

CHAPTER 6. DIFFERENTIABILITY

which implies
$$\lim_{x \to c+} \frac{f(x) - f(c)}{x - c} = l .$$

Similarly we will get
$$\lim_{x \to c-} \frac{f(x) - f(c)}{x - c} = l ,$$

which implies $f(x)$ is differentiable at c and $f'(c) = l$.

$\boxed{\text{Solution 6.17}}$

Assume $h > 0$. The Mean Value Theorem implies the existence of $c \in (x_0, x_0 + h)$ such that
$$\frac{f(x_0 + h) - f(x_0)}{h} = f'(c) .$$

Set $\theta = \frac{(c - x_0)}{h}$. Then $c = x_0 + \theta h$ and $\theta \in (0, 1)$. Hence
$$\frac{f(x_0 + h) - f(x_0)}{h} = f'(x_0 + \theta h)$$

or $f(x_0 + h) - f(x_0) = h f'(x_0 + \theta h)$. When $h < 0$, similar ideas will lead to the same conclusion. If $f(x) = \frac{1}{1+x}$, then we must have
$$\frac{1}{1 + x_0 + h} - \frac{1}{1 + x_0} = h \left(-\frac{1}{(1 + x_0 + \theta h)^2} \right) = -\frac{h}{(1 + x_0 + \theta h)^2} .$$

Easy algebraic manipulations will then imply
$$-\frac{h}{(1 + x_0)(1 + x_0 + h)} = -\frac{h}{(1 + x_0 + \theta h)^2}$$

which implies $(1 + x_0 + \theta h)^2 = (1 + x_0)(1 + x_0 + h)$. Since $h \to 0$, we may assume that $(1 + x_0 + h) > 0$ and $1 + x_0 + \theta h > 0$. Hence
$$\theta = \frac{\sqrt{(1 + x_0)(1 + x_0 + h)} - 1 - x_0}{h} .$$

Since
$$\frac{\sqrt{(1 + x_0)(1 + x_0 + h)} - 1 - x_0}{h} = \frac{(1 + x_0)(1 + x_0 + h) - (1 + x_0)^2}{h \left(\sqrt{(1 + x_0)(1 + x_0 + h)} + 1 + x_0 \right)}$$

or
$$\frac{\sqrt{(1 + x_0)(1 + x_0 + h)} - 1 - x_0}{h} = \frac{h(1 + x_0)}{h \left(\sqrt{(1 + x_0)(1 + x_0 + h)} + 1 + x_0 \right)} ,$$

$$\theta = \frac{(1 + x_0)}{\left(\sqrt{(1 + x_0)(1 + x_0 + h)} + 1 + x_0 \right)} .$$

Hence
$$\lim_{h \to 0} \theta = \frac{(1 + x_0)}{\left(\sqrt{(1 + x_0)(1 + x_0)} + 1 + x_0 \right)} = \frac{1}{2} .$$

Solution 6.18

Let us start by proving (a). First assume $x > 0$. The Mean Value Theorem applied to $\ln(1+x)$ on the interval $[0, x]$ implies the existence of $c \in (0, x)$ such that

$$\frac{\ln(1+x) - \ln(1+0)}{x - 0} = \frac{1}{1+c},$$

or $\ln(1+x) = \dfrac{x}{1+c}$. Since $0 < \dfrac{1}{1+c} < 1$, we get $\ln(1+x) < x$. This implies $\ln(1+x) \leq x$ for any $x \geq 0$.

For (b), set $f(x) = \tan(x) - x - \dfrac{x^3}{3}$. Let $x > 0$. Let us apply the Mean Value Theorem to $f(x)$ on $[0, x]$. Then there exists $c \in (0, x)$ such that

$$\frac{f(x) - f(0)}{x - 0} = f'(c).$$

Since

$$\frac{f(x) - f(0)}{x - 0} = \frac{\tan(x)}{x} - 1 - \frac{x^2}{3} \quad \text{and} \quad f'(c) = 1 + \tan^2(c) - 1 - c^2 = \tan^2(c) - c^2,$$

we get

$$\tan(x) - x - \frac{x^3}{3} = x \left(\tan^2(c) - c^2 \right).$$

The proof will be complete if we prove that $\tan(x) - x > 0$ for any $x \in \left(0, \dfrac{\pi}{2}\right)$. In order to get it done, let us apply the Mean Value Theorem to $\tan(x)$ on the interval $[0, x]$. So there exists $c^* \in (0, x)$ such that

$$\frac{\tan(x) - \tan(0)}{x - 0} = \sec^2(c) = 1 + \tan^2(c).$$

Since $1 + \tan^2(c) > 1$, then we have $\dfrac{\tan(x)}{x} > 1$ or $x < \tan(x)$. Therefore we have

$$\left(\tan^2(c) - c^2 \right) = \left(\tan(c) - c \right) \left(\tan(c) + c \right) > 0$$

which implies $\tan(x) - x - \dfrac{x^3}{3} > 0$ for any $x \in \left(0, \dfrac{\pi}{2}\right)$.

Since $\sin'(x) = \cos(x)$ and $\cos'(x) = -\sin(x)$, we will prove (c) and (d) simultaneously. First for any $x \in (0, \dfrac{\pi}{2})$, the Mean Value Theorem implies the existence of $c \in (0, x)$ such that

$$\frac{\sin(x) - \sin(0)}{x - 0} = \frac{\sin(x)}{x} = \cos(c) < 1$$

which implies $\sin(x) < x$. Now set

$$g(x) = 1 - \frac{x^2}{2} - \cos(x).$$

When we apply the Mean Value Theorem to $g(x)$ on $(0, x)$ we secure the existence of $d \in (0, x)$ such that

$$\frac{g(x) - g(0)}{x - 0} = \frac{1 - \frac{x^2}{2} - \cos(x)}{x} = g'(d) = \sin(d) - d < 0$$

from the previous argument. Since $x > 0$ we get

$$\cos(x) > 1 - \frac{x^2}{2}.$$

In order to show that $x - \frac{x^3}{6} < \sin(x)$ and $\cos(x) < 1 - \frac{x^2}{2} + \frac{x^4}{24}$, apply the Mean Value Theorem to the functions

$$h(x) = \sin(x) - x + \frac{x^3}{6} \quad \text{and} \quad k(x) = 1 - \frac{x^2}{2} + \frac{x^4}{24} - \cos(x)$$

for $0 < x \leq \frac{\pi}{2}$.

Solution 6.19

Assume that $f(x)$ is not one-to-one. Then there exists $x_1 < x_2$ such that $f(x_1) = f(x_2)$. Rolle's theorem will imply the existence of $c \in (x_1, x_2)$ such that $f'(c) = 0$ which contradicts our assumption. Next we will prove that $f(x)$ is monotone. Without loss of generality, assume $f(a) < f(b)$. Let $x \in (a, b)$. Assume $f(a) < f(b) < f(x)$. Since $f(x)$ is continuous on $[a, b]$, the Intermediate Value Theorem implies the existence of $c \in (a, x)$ such that $f(c) = f(b)$. Clearly $c \neq b$ which generates a contradiction with $f(x)$ being one-to-one. The same ideas will imply that $f(x) < f(a) < f(b)$ does not hold. Therefore we must have $f(a) < f(x) < f(b)$ for any $x \in (a, b)$. Next let $x, y \in (a, b)$ such that $x < y$. Assume $f(y) < f(x)$. Then we have $f(y) < f(x) < f(b)$. Again the Intermediate Value Theorem implies the existence of $c \in (y, b)$ such that $f(c) = f(x)$. Clearly $c \neq x$ because $x < y$. This is a contradiction with $f(x)$ being one-to-one. Therefore, for any $x, y \in (a, b)$ with $x < y$, we have $f(x) < f(y)$, i.e., $f(x)$ is increasing. This will imply that $f'(x) \geq 0$ but since $f'(x) \neq 0$, we get $f'(x) > 0$ for any $x \in (a, b)$. Finally let us prove that $f'(x)$ satisfies the conclusion of the Intermediate Value Theorem. Indeed, let $x_1, x_2 \in (a, b)$ and $\alpha \in \mathbb{R}$ such that $f'(x_1) < \alpha < f'(x_2)$. Without loss of generality we assume that $x_1 < x_2$. Next define $g(x) = f(x) - \alpha x$. The function $g(x)$ inherits all the properties of $f(x)$. In particular, we have $g'(x) = f'(x) - \alpha$. Assume that $g'(x) \neq 0$ for any $x \in (x_1, x_2)$. From the first part, we deduce that $g'(x) > 0$ or $g'(x) < 0$ for any $x \in (x_1, x_2)$. It is easy to check that this conclusion still holds at x_1 and x_2. Hence $g'(x) > 0$ or $g'(x) < 0$ for any $x \in [x_1, x_2]$. In other words, we have $f'(x) < \alpha$ or $f'(x) > \alpha$ for any $x \in [x_1, x_2]$. This is a contradiction with $f'(x_1) < \alpha < f'(x_2)$. Therefore there exists $c \in (x_1, x_2)$ such that $g'(c) = 0$ or $f'(c) = \alpha$. This completes the proof of our statement.

Solution 6.20

Set $g(x) = e^{-x} f(x)$. Then $g(x)$ is differentiable and $g'(x) = e^{-x}\Big(f'(x) - f(x)\Big)$. Our assumption implies $g'(x) > 0$ which leads to $g(x)$ being increasing. Since $g(x_0) = 0$, $g(x) > 0$ for $x > x_0$. Since $f(x) = e^x g(x)$, we get $f(x) > 0$ for $x > x_0$. For the application, define the function

$$f(x) = ae^x - 1 - x - \frac{x^2}{2}.$$

Note that
$$f'(x) = ae^x - 1 - x > ae^x - 1 - x - \frac{x^2}{2} = f(x).$$
Since
$$\lim_{x \to \infty} f(x) = \infty \quad \text{and} \quad \lim_{x \to -\infty} f(x) = -\infty,$$
there exists x_0 such that $f(x_0) = 0$. This implies the existence of at least one root to the equation. Let us prove that there is only one root. Assume not. Let $x_0 < x_1$ be two different roots. The previous argument shows that $f(x) > 0$ for $x > x_0$ which contradicts $f(x_1) = 0$. This completes the proof of our statements.

Solution 6.21

For any $x \neq t$, we have
$$\frac{|f(x) - f(t)|}{|x - t|} < |x - t|^\alpha.$$
Since $\lim_{x \to t} |x - t|^\alpha = 0$, we deduce that
$$\lim_{x \to t} \left| \frac{f(x) - f(t)}{x - t} \right| = \lim_{x \to t} \frac{|f(x) - f(t)|}{|x - t|} = 0.$$
Hence
$$\lim_{x \to t} \frac{f(x) - f(t)}{x - t} = 0,$$
therefore $f'(t)$ exists for any $t \in \mathbb{R}$. Since
$$f'(t) = \lim_{x \to t} \frac{f(x) - f(t)}{x - t} = 0,$$
we deduce that $f(x)$ is constant.

Solution 6.22

Set $g(x) = e^x f(x)$. Then we have $g'(x) = e^x \Big(f(x) + f'(x) \Big)$. Let $\varepsilon > 0$. There exists $A > 0$ such that for any $x > A$ we have $|f(x) + f'(x)| < \frac{\varepsilon}{2}$. Let $x > A$, the generalized Mean Value Theorem implies the existence of $c \in (A, x)$ such that
$$\frac{g'(c)}{e^c} = \frac{g(x) - g(A)}{e^x - e^A},$$
or
$$f(c) + f'(c) = \frac{g(x) - g(A)}{e^x - e^A}.$$
In particular, we have $|g(x) - g(A)| < \frac{\varepsilon}{2} |e^x - e^A|$, which implies $|g(x)| < \frac{\varepsilon}{2} |e^x - e^A| + |g(A)|$ or
$$|f(x)| < \frac{\varepsilon}{2} |1 - e^{A-x}| + |f(A) e^{A-x}| < \frac{\varepsilon}{2} + |f(A) e^{A-x}|$$

CHAPTER 6. DIFFERENTIABILITY

because $0 < e^{A-x} < 1$. Since $\lim_{x \to \infty} |f(A)e^{A-x}| = 0$, there exists $B > 0$ such that for any $x > B$ we have $|f(A)e^{A-x}| < \dfrac{\varepsilon}{2}$. Let $A^* = \max\{A, B\}$, then for any $x > A^*$ we have

$$|f(x)| < \frac{\varepsilon}{2} + \frac{\varepsilon}{2} = \varepsilon \,.$$

This completes the proof of our statement.

Solution 6.23

Let $D = \{x \in [0,1]; f(t) = 0 \text{ for } t \in [0,x]\}$. Since $0 \in D$, D is a nonempty subset of $[0,1]$. So the supremum of D exists. Set $s = \sup D$. Note that continuity of $f(x)$ implies $s \in D$. In order to complete the proof of our statement, we want to show that $s = 1$. Assume otherwise that $s < 1$. Then there exists $a_0 > 0$ such that $s + a_0 < 1$ and $a_0 M < 1$. For any $x \in (s, s + a_0)$, the Mean Value Theorem ensures the existence of $c \in (s,x)$ such that

$$f(x) - f(s) = f'(c)(x-s) \,.$$

Since $f(s) = 0$, we get $f(x) = f'(c)(x-s)$. Hence

$$|f(x)| \leq |f'(c)||x-s| \leq M a_0 \max_{s \leq t \leq s+a_0} |f(t)|$$

for any $x \in [s, s+a_0]$. Hence

$$\max_{s \leq x \leq s+a_0} |f(x)| \leq a_0 M \max_{s \leq t \leq s+a_0} |f(t)| \,.$$

Since $a_0 M < 1$, we get

$$\max_{s \leq x \leq s+a_0} |f(x)| < \max_{s \leq t \leq s+a_0} |f(t)|$$

which is the desired contradiction. So $s = 1$ or $f(x) = 0$ for any $x \in [0,1]$.

Solution 6.24

First note that for $x \neq 0$, we have

$$f'(x) = \frac{2}{x^3} e^{-1/x^2} \,.$$

Consider the polynomial function $P_1(x) = 2x^3$, then we have

$$f'(x) = P_1\left(\frac{1}{x}\right) e^{-1/x^2} \,.$$

Since

$$\lim_{x \to 0} \frac{1}{x^n} e^{-1/x^2} = 0 \,,$$

for any natural integer $n \geq 0$,

$$\lim_{x \to 0} P\left(\frac{1}{x}\right) e^{-1/x^2} = 0 \,,$$

for any polynomial function $P(x)$. Hence $\lim_{x \to 0} f'(x) = 0$. By Problem 6.16, we conclude that $f(x)$ is differentiable at 0 and $f'(0) = 0$. By induction, we prove the existence of a polynomial function $P_n(x)$ such that

$$f^{(n)}(x) = P_n\left(\frac{1}{x}\right) e^{-1/x^2},$$

for any $x \neq 0$. Indeed, the previous calculations show that the induction conclusion is true for $n = 1$. Assume it is true for $n = k$, let us prove it is still true for $n = k+1$. We have

$$f^{(k+1)}(x) = \left(f^{(k)}(x)\right)' = \left(P_k\left(\frac{1}{x}\right) e^{-1/x^2}\right)'$$

which implies

$$f^{(k+1)}(x) = \left(-\frac{1}{x^2} P'_k\left(\frac{1}{x}\right) + \frac{2}{x^3} P_k\left(\frac{1}{x}\right)\right) e^{-1/x^2}.$$

Set

$$P_{k+1}(x) = -x^2 P'_k(x) + 2x^3 P_k(x).$$

Then we have

$$f^{(k+1)}(x) = P_{k+1}\left(\frac{1}{x}\right) e^{-1/x^2},$$

which shows that the claim is true for $n = k+1$. This completes the proof of the induction. From this conclusion we get $\lim_{x \to 0} f^{(n)}(x) = 0$ for any $n \geq 0$. By Problem 6.16, we conclude that $f^{(n)}(x)$ is differentiable at 0 and $f^{(n+1)}(0) = 0$ for any $n \geq 0$.

Solution 6.25

Fix $x, c \in (a, b)$ with $x \neq c$. Set

$$F(t) = f(t) + f'(t)(x - t) + M(x - t)^2$$

where M is chosen such that $F(c) = f(x)$. Then we have $F(x) = f(x) = F(c)$. The Mean Value Theorem implies the existence of θ between x and c such that $F'(\theta) = 0$. But

$$F'(\theta) = f'(\theta) + f''(\theta)(x - \theta) - f'(\theta) - 2M(x - \theta) = 0$$

which implies

$$M = \frac{f''(\theta)}{2}.$$

So for $x, c \in (a, b)$, there exists θ between x and c such that

$$f(x) = f(c) + f'(c)(x - c) + \frac{f''(\theta)}{2}(x - c)^2.$$

So for any $h > 0$, there exist $\theta_1 \in (c, c + h)$ and $\theta_1 \in (c - h, c)$ such that

$$f(c + h) = f(c) + f'(c)h + \frac{f''(\theta_1)}{2} h^2$$

and

$$f(c - h) = f(c) - f'(c)h + \frac{f''(\theta_2)}{2} h^2.$$

CHAPTER 6. DIFFERENTIABILITY

So
$$\frac{f(c+h) - 2f(c) + f(c-h)}{h^2} = \frac{f''(\theta_1) + f''(\theta_2)}{2}.$$

It is clear that when $h \to 0$, then $\theta_i \to c$ for $i = 1, 2$. And since $f''(x)$ is continuous at c, we get
$$\lim_{h \to 0} \frac{f(c+h) - 2f(c) + f(c-h)}{h^2} = f''(c).$$

For the converse, the answer is in the negative. Indeed, take $f(x) = x|x|$ and $c = 0$. Then we have
$$\lim_{h \to 0} \frac{f(0+h) - 2f(c) + f(0-h)}{h^2} = 0$$

but $f''(0)$ does not exist.

Solution 6.26

Since $f(\tilde{x}) = 0$, the Mean Value Theorem gives
$$|f(x)| = |f(x) - f(\tilde{x})| = |f'(z)| \cdot |x - \tilde{x}| \le k_1 |x - \tilde{x}|$$

with $k_1 > 0$. Since $\tilde{x}$ is simple, $f'(x) \ne 0$ on some neighborhood U of $\tilde{x}$, $U \subset [a,b]$. f'' is bounded on U, and for any $x \in U$,
$$|g'(x)| = \left|1 - \frac{f'(x)f'(x) - f''(x)f(x)}{(f'(x))^2}\right| = \left|1 - 1 - \frac{f''(x)f(x)}{(f'(x))^2}\right|$$
$$\le k_2 |f(x)| \le k_1 k_2 |x - \tilde{x}| < \frac{1}{2}$$

whenever $|x - \tilde{x}| < 1/2k_1 k_2$, thus g is a contraction, and by the Banach Contraction Mapping Theorem iterative sequence converges to $\tilde{x}$.

Solution 6.27

Let $a \in (0, 1)$. The proof of the previous problem or Taylor expansion with remainder of $f(x)$ at a will imply
$$f(x) = f(a) + f'(a)(x - a) + \frac{f''(c)}{2}(x - a)^2$$

where c is between a and x. Setting $x = 0$ and $x = 1$ in the above equation results in
$$f(0) = f(a) + f'(a)(0 - a) + \frac{f''(c_1)}{2}(0 - a)^2$$

and
$$f(1) = f(a) + f'(a)(1 - a) + \frac{f''(c_2)}{2}(1 - a)^2$$

where $0 < c_1 < a$ and $a < c_2 < 1$. Subtract the first equation from the second to get
$$0 = f'(a) + \frac{f''(c_2)}{2}(1 - a)^2 - \frac{f''(c_1)}{2}a^2.$$

Hence
$$f'(a) = \frac{f''(c_1)}{2}a^2 - \frac{f''(c_2)}{2}(1 - a)^2.$$

If we use the fact $|f''(x)| \leq A$, we get

$$|f'(a)| \leq \frac{A}{2}\left(a^2 + (1-a)^2\right).$$

Set $a = \frac{1}{2}$ to get

$$\left|f'\left(\frac{1}{2}\right)\right| \leq \frac{A}{4}.$$

Also if we note that $a^2 + (1-a)^2 \leq 1$, we get

$$|f'(a)| \leq \frac{A}{2}.$$

Solution 6.28

We first give Lagrange's version. Define F by setting

$$F(x) = f(a+h) - f(x) - (a+h-x)f'(x) - \frac{(a+h-x)^2}{2!}f''(x) - \cdots - \frac{(a+h-x)^{n-1}}{(n-1)!}f^{(n-1)}(x)$$

$$= f(a+h) - \sum_{k=0}^{n-1} \frac{(a+h-x)^k}{k!} f^{(k)}(x),$$

and $G(x) = F(x) - \left(\frac{a+h-x}{h}\right)^n F(a)$.

Then the hypotheses imply that the functions F and G are differentiable on I. Also,

$$G'(x) = F'(x) + \frac{n}{h}\left(\frac{a+h-x}{h}\right)^{n-1} F(a)$$

$$= \sum_{k=1}^{n-1} \frac{(a+h-x)^{k-1}}{(k-1)!} f^{(k)}(x) - \sum_{k=0}^{n-1} \frac{(a+h-x)^k}{k!} f^{(k+1)}(x) + \frac{n}{h}\left(\frac{a+h-x}{h}\right)^{n-1} F(a)$$

$$= \sum_{k=0}^{n-2} \frac{(a+h-x)^k}{k!} f^{(k+1)}(x) - \sum_{k=0}^{n-1} \frac{(a+h-x)^k}{k!} f^{(k+1)}(x) + \frac{n}{h}\left(\frac{a+h-x}{h}\right)^{n-1} F(a)$$

$$= -\frac{(a+h-x)^{n-1}}{(n-1)!} f^{(n)}(x) + \frac{n}{h}\left(\frac{a+h-x}{h}\right)^{n-1} F(a).$$

Now $G(a) = F(a) - \left(\frac{a+h-a}{h}\right)^n F(a) = 0$ and

$$G(a+h) = F(a+h) - \left(\frac{a+h-a-h}{h}\right)^n F(a) = F(a+h) = f(a+h) - f(a+h) = 0.$$

So, by Rolle's Theorem, there exists c between a and $a+h$ such that $G'(c) = 0$. Write c as $a + \theta h$ where $0 < \theta < 1$. Thus

$$-\frac{(a+h-a-\theta h)^{n-1}}{(n-1)!} f^{(n)}(a+\theta h) + \frac{n}{h}\left(\frac{a+h-a-\theta h}{h}\right)^{n-1} F(a) = 0.$$

CHAPTER 6. DIFFERENTIABILITY   121

Now $a + h - a - \theta h = (1-\theta)h \neq 0$, so this gives $F(a) = \dfrac{h^n}{n!}f^{(n)}(a+\theta h) = R_n$.

Hence $f(a+h) = \displaystyle\sum_{k=0}^{n-1}\dfrac{h^k}{k!}f^{(k)}(a) + R_n$, as required.

The proof for the Cauchy form of the remainder is similar, but a bit simpler, using the Mean Value Theorem instead of Rolle's Theorem.

Let $F(x) = \displaystyle\sum_{k=0}^{n-1}\dfrac{(a+h-x)^k}{k!}f^{(k)}(x)$. By the Mean Value Theorem, there exists $\theta \in (0,1)$ such that $F(a+h) = F(a) + hF'(a+\theta h)$.

Hence $F(a+h) = \displaystyle\sum_{k=0}^{n-1}\dfrac{h^k}{k!}f^{(k)}(a) + hF'(a+\theta h)$. But

$$F'(x) = -\sum_{k=1}^{n-1}\dfrac{(a+h-x)^{k-1}}{(k-1)!}f^{(k)}(x) + \sum_{k=0}^{n-1}\dfrac{(a+h-x)^k}{k!}f^{(k+1)}(x)$$
$$= -\sum_{k=0}^{n-2}\dfrac{(a+h-x)^k}{k!}f^{(k+1)}(x) + \sum_{k=0}^{n-1}\dfrac{(a+h-x)^k}{k!}f^{(k+1)}(x)$$
$$= \dfrac{(a+h-x)^{n-1}}{(n-1)!}f^{(n)}(x),$$

so $F'(a+\theta h) = \dfrac{(a+h-a-\theta h)^{n-1}}{(n-1)!}f^{(n)}(a+\theta h) = \dfrac{h^{n-1}(1-\theta)^{n-1}}{(n-1)!}f^{(n)}(a+\theta h)$, which gives the result.

Solution 6.29

We just prove (ii). First observe that the given series is indeed the Maclaurin series of $f(x) = (1+x)^s$, since $f'(x) = s(1+x)^{s-1}$, $f''(x) = s(s-1)(1+x)^s$, from which $f(0) = 1$, $f'(0) = s$, $f''(0) = s(s-1)$, $f'''(0) = s(s-1)(s-2)$, etc. So the main point is to show that the remainder $R_n \to 0$ as $n \to \infty$.

We use the Cauchy form for R_n,

$$R_n = \dfrac{x^n(1-\theta)^{n-1}}{(n-1)!}f^{(n)}(\theta x)$$
$$= \dfrac{x^n(1-\theta)^{n-1}s(s-1)\cdots(s-n+1)(1+\theta x)^{s-n}}{(n-1)!}$$
$$= (1+\theta x)^{s-1} \times \left(\dfrac{1-\theta}{1+\theta x}\right)^{n-1} \times \dfrac{s(s-1)\cdots(s-n+1)x^n}{(n-1)!}$$
$$= \quad\ a_n \quad\ \times \quad\ b_n \quad\ \times \quad\ c_n,$$

say. We now estimate $|a_n|$, $|b_n|$, and $|c_n|$.

For a_n, notice that if $s > 1$, then $|a_n| \leq (1+|x|)^{s-1}$ and that if $s < 1$, then $|a_n| \leq (1-|x|)^{s-1}$. In either case, $|a_n|$ is bounded independently of n, say $|a_n| \leq K$.

For b_n, notice that $1 + \theta x > 1 - \theta$, and so $|b_n| < 1$.

For c_n, we have
$$\left|\frac{c_{n+1}}{c_n}\right| = \left|\frac{s-n}{n}x\right| \to |x| \text{ as } n \to \infty.$$

Fix l with $|x| < l < 1$. Then there exists $N \in \mathbb{N}$ such that $|c_{n+1}/c_n| \le l$ for all $n \ge N$. It follows that $|c_n| \le l^{n-N}|c_N|$ whenever $n \ge N$. Hence $c_n \to 0$ as $n \to \infty$.

Putting this all together, we obtain
$$|R_n| = |a_n| \cdot |b_n| \cdot |c_n| \le K \cdot 1 \cdot |c_n| \to 0 \text{ as } n \to \infty.$$

Solution 6.30

By induction, the conclusion is obvious when $n = 0$. For $n = 1$, the formula reduces to the well-known product formula for differentiation. Assume that
$$\frac{d^{n-1}}{dx^{n-1}}(fg) = \sum_{k=0}^{n-1} \binom{n-1}{k} f^{[k]} g^{[n-1-k]},$$

where $f^{[k]} = \frac{d^k}{dx^k} f$, for any $k \in \mathbb{N}$. Then
$$\frac{d^n}{dx^n}(fg) = \frac{d}{dx}\left(\frac{d^{n-1}}{dx^{n-1}}(fg)\right) = \sum_{k=0}^{n-1} \binom{n-1}{k} \left(f^{[k]} g^{[n-k]} + f^{[k+1]} g^{[n-1-k]}\right)$$

which implies

$$\begin{aligned}
\frac{d^n}{dx^n}(fg) &= fg^{[n]} + \sum_{k=1}^{n-1} \binom{n-1}{k} f^{[k]} g^{[n-k]} + \sum_{k=0}^{n-2} \binom{n-1}{k} f^{[k+1]} g^{[n-1-k]} + f^{[n]} g \\
&= fg^{[n]} + \sum_{k=1}^{n-1} \binom{n-1}{k} f^{[k]} g^{[n-k]} + \sum_{k=1}^{n-1} \binom{n-1}{k-1} f^{[k]} g^{[n-k]} + f^{[n]} g \\
&= fg^{[n]} + \sum_{k=1}^{n-1} \left(\binom{n-1}{k} + \binom{n-1}{k-1}\right) f^{[k]} g^{[n-k]} + f^{[n]} g \\
&= fg^{[n]} + \sum_{k=1}^{n-1} \binom{n}{k} f^{[k]} g^{[n-k]} + f^{[n]} g \\
&= \sum_{k=0}^{n} \binom{n}{k} f^{[k]} g^{[n-k]}.
\end{aligned}$$

CHAPTER 6. DIFFERENTIABILITY

Note that we made use of the formula
$$\binom{n-1}{k} + \binom{n-1}{k-1} = \binom{n}{k}.$$

This completes the proof by induction on $n \in \mathbb{N}$.

Solution 6.31

Let
$$T : (C[a,b], \| \cdot \|) \to (C[a,b], \| \cdot \|)$$
defined as
$$Ty(x) = y(x) - \frac{1}{M} F(x, y(x)).$$

We first claim that T is a contraction on $C[a,b]$. From the Mean Value Theorem we have
$$F(x, y_1) - F(x, y_2) = \frac{\partial F}{\partial y}(x, c)(y_1 - y_2)$$

where c is between y_1 and y_2. Hence

$$T_{y_1}(x) - T_{y_2}(x) = [y_1(x) - y_2(x)] - \frac{1}{M}[F(x, y_1) - F(x, y_2)] = [y_1(x) - y_2(x)] - \frac{1}{M}\frac{\partial F}{\partial y}(x, c)[y_1(x) - y_2(x)]$$

$$= \left(1 - \frac{1}{M}\frac{\partial F}{\partial y}(x, c)\right)[y_1(x) - y_2(x)] \le \left(1 - \frac{m}{M}\right)[y_1(x) - y_2(x)].$$

Consider the given norm $\| \cdot \|$,

$$\| Ty_1 - Ty_2 \| = \max_{a \le x \le b} |Ty_1(x) - Ty_2(x)| \le \max_{a \le x \le b} \left(1 - \frac{m}{M}\right)|y_1(x) - y_2(x)| \le \left|1 - \frac{m}{M}\right| \| y_1 - y_2 \|.$$

Since $1 - \dfrac{m}{M} < 1$ it follows that T is a contraction on $C[a,b]$, therefore by the Banach Contraction Mapping Theorem T has a unique fixed point y. Thus for all $x \in [a,b]$,

$$y(x) = Ty(x) = y(x) - \frac{1}{M} F(x, y(x))$$

but $M \ne 0$ thus $F(x, y(x)) = 0$.

Solution 6.32

1. Using the Leibnitz formula (see Problem 6.30), we get

$$\frac{1}{2^n n!} \frac{d^n}{dx^n}\left(x(x^2 - 1)^n\right) = x \frac{1}{2^n n!} \frac{d^n}{dx^n}\left((x^2 - 1)^n\right) + n \frac{1}{2^n n!} \frac{d^{n-1}}{dx^{n-1}}\left(x(x^2 - 1)^n\right)$$

which implies

$$xL_n = \frac{1}{2^n n!} \frac{d^n}{dx^n}\left(x(x^2 - 1)^n\right) - \frac{n}{2^n n!} \frac{d^{n-1}}{dx^{n-1}}\left((x^2 - 1)^n\right).$$

But
$$\frac{1}{2^n n!} \frac{d^n}{dx^n}\left(x(x^2-1)^n\right) = \frac{1}{2^n n!} \frac{d^{n+1}}{dx^{n+1}}\left(\frac{1}{2(n+1)}(x^2-1)^{n+1}\right),$$
or
$$\frac{1}{2^n n!} \frac{d^n}{dx^n}\left(x(x^2-1)^n\right) = \frac{1}{2^{n+1}(n+1)!} \frac{d^{n+1}}{dx^{n+1}}\left((x^2-1)^{n+1}\right) = L_{n+1}.$$
Hence
$$xL_n = L_{n+1} - \frac{n}{2^n n!} \frac{d^{n-1}}{dx^{n-1}}\left((x^2-1)^n\right).$$
On the other hand, we have
$$L_n = \frac{1}{2^n n!} \frac{d^n}{dx^n}\left((x^2-1)^n\right) = \frac{1}{2^n n!} \frac{d^{n-1}}{dx^{n-1}}\left(2nx(x^2-1)^n\right),$$
which implies
$$xL_n = \frac{x}{2^{n-1}(n-1)!} \frac{d^{n-1}}{dx^{n-1}}\left(x(x^2-1)^{n-1}\right).$$
Using the Leibnitz formula again we get $\left((x^2-1)^n\right)^{[n-1]} = 2n\left(x(x^2-1)^{n-1}\right)^{[n-2]}$, and
$$\left(x^2(x^2-1)^{n-1}\right)^{[n-1]} = x\left(x(x^2-1)^{n-1}\right)^{[n-1]} + (n-1)\left(x(x^2-1)^{n-1}\right)^{[n-2]}.$$
Hence
$$\frac{1}{2^{n-1}(n-1)!}\left(x^2(x^2-1)^{n-1}\right)^{[n-1]} = \frac{x}{2^{n-1}(n-1)!}\left(x(x^2-1)^{n-1}\right)^{[n-1]} + \frac{(n-1)}{2^n(n)!}\left((x^2-1)^n\right)^{[n-1]},$$
which implies
$$xL_n = \frac{1}{2^{n-1}(n-1)!}\left(x^2(x^2-1)^{n-1}\right)^{[n-1]} - \frac{(n-1)}{2^n(n)!}\left((x^2-1)^n\right)^{[n-1]}.$$
But
$$\left(x^2(x^2-1)^{n-1}\right)^{[n-1]} = \left((x^2-1)(x^2-1)^{n-1}\right)^{[n-1]} + \left((x^2-1)^{n-1}\right)^{[n-1]}$$
which combined with the previous identity gives
$$xL_n = \frac{(n+1)}{2^n n!}\left((x^2-1)^n\right)^{[n-1]} + \frac{1}{2^{n-1}(n-1)!}\left((x^2-1)^{n-1}\right)^{[n-1]},$$
or
$$xL_n = \frac{(n+1)}{2^n n!}\left((x^2-1)^n\right)^{[n-1]} + L_{n-1}.$$
Hence we have
$$\frac{x}{n} L_n = \frac{1}{n} L_{n+1} - \frac{1}{2^n n!}\left((x^2-1)^n\right)^{[n-1]}$$
and
$$\frac{x}{n+1} L_n = \frac{1}{n+1} L_{n-1} + \frac{1}{2^n n!}\left((x^2-1)^n\right)^{[n-1]}.$$
So if we add both equations we get
$$(2n+1)xL_n = (n+1)L_{n+1} + nL_{n-1},$$
which gives the desired recurrence formula.

CHAPTER 6. DIFFERENTIABILITY

2. Note that we have

$$L_0(x) = 1, \; L_1(x) = \frac{1}{2}\Big(2x\Big) = x, \; L_2(x) = -\frac{3x^2}{2} - \frac{1}{2}, \; L_3(x) = -\frac{3x}{2} + \frac{5x^3}{2}.$$

From this we claim that $L_n(x)$ has degree n for all $n \in \mathbb{N}$. Indeed the claim is true for $n = 0, 1, 2, 3$. Hence assume that L_k has degree k, for any $k \leq n$. Let us prove that L_{n+1} has degree $n+1$. The recurrence formula proved above gives

$$L_{n+1} = \frac{(2n+1)}{(n+1)}xL_n - \frac{n}{n+1}L_{n-1}.$$

Since L_{n-1} has degree $n-1$ and xL_n has degree $n+1$, we conclude that L_{n+1} has degree $n+1$. By induction we conclude that L_n has degree n for any $n \in \mathbb{N}$.

3. First note that for any $k \in \mathbb{N}$ and $n > k$, we have

$$\frac{d^k}{dx^k}\Big((x^2-1)^n\Big) = (x^2-1)^{n-k} P(x),$$

for some polynomial function $P(x)$. Hence for any differentiable function $f(x)$ defined on $[-1, 1]$, we have

$$\int_{-1}^{1} f(x) L_n(x) dx = \frac{1}{2^n n!} \int_{-1}^{1} f(x)\Big((x^2-1)^n\Big)^{[n]} dx.$$

Integrating by parts, we get

$$\int_{-1}^{1} f(x) L_n(x) dx = \frac{1}{2^n n!} \left[f(x)\Big((x^2-1)^n\Big)^{[n-1]} \right]_{-1}^{1} - \frac{1}{2^n n!} \int_{-1}^{1} f'(x)\Big((x^2-1)^n\Big)^{[n-1]} dx.$$

Using the above remark on the derivatives of $(x^2-1)^n$, we get

$$\int_{-1}^{1} f(x) L_n(x) dx = \frac{-1}{2^n n!} \int_{-1}^{1} f'(x)\Big((x^2-1)^n\Big)^{[n-1]} dx,$$

which implies

$$\int_{-1}^{1} f(x) L_n(x) dx = \frac{(-1)^n}{2^n n!} \int_{-1}^{1} f^{[n]}(x)(x^2-1)^n dx.$$

In particular, we have

$$\int_{-1}^{1} x^k L_n(x) dx = \frac{(-1)^n}{2^n n!} \int_{-1}^{1} (x^k)^{[n]}(x^2-1)^n dx = 0,$$

since $(x^k)^{[n]} = 0$, if $k < n$. This clearly implies

$$\int_{-1}^{1} P(x) L_n(x) dx = 0,$$

for any polynomial function $P(x)$ with degree less than n. Hence

$$\int_{-1}^{1} L_m(x) L_n(x) dx = 0,$$

for any $n \neq m$.

4. We have

$$\int_{-1}^{1}\left(L_n(x)\right)^2 dx = \frac{(-1)^n}{2^n n!}\int_{-1}^{1} L_n^{[n]}(x)(x^2-1)^n dx = \frac{(-1)^n}{2^{2n}(n!)^2}\int_{-1}^{1}(x^2-1)^n\left((x^2-1)^n\right)^{[2n]} dx.$$

Note that $\left((x^2-1)^n\right)^{[2n]} = (2n)!$, since the polynomial function $(x^2-1)^n$ has degree $2n$ with the leading coefficient being equal to 1. So

$$\int_{-1}^{1}\left(L_n(x)\right)^2 dx = \frac{(-1)^n(2n)!}{2^{2n}(n!)^2}\int_{-1}^{1}(x^2-1)^n dx = 2\frac{(2n)!}{2^{2n}(n!)^2}\int_{0}^{1}(1-x^2)^n dx$$

since the function $(1-x^2)^n$ is even. If we do the change of variable $x = \cos(t)$, we get

$$\int_{-1}^{1}\left(L_n(x)\right)^2 dx = 2\frac{(2n)!}{2^{2n}(n!)^2}\int_{0}^{\pi/2}\cos^{2n+1}(t)dt.$$

We recognize Wallis integrals (see Problem 3.21). Since

$$\int_{0}^{\pi/2}\cos^{2n+1}(t)dt = \frac{2^{2n}(n!)^2}{(2n+1)!},$$

$$\int_{-1}^{1}\left(L_n(x)\right)^2 dx = 2\frac{(2n)!}{2^{2n}(n!)^2}\frac{2^{2n}(n!)^2}{(2n+1)!} = \frac{2}{2n+1}.$$

Chapter 7

Integration

Nature laughs at the difficulties of integration.

Pierre-Simon Laplace (1749–1827)

- *Partitions, Upper and Lower Sums*: Let f be a bounded function on $[a,b]$. A *partition* of $[a,b]$ is a finite ordered set

$$P = \{a = x_0 < x_1 < x_2 < \ldots < x_n = b\}.$$

For each subinterval $[x_{k-1}, x_k]$ of P set

$$m_k = \inf\{f(x) : \ x \in [x_{k-1}, x_k]\}$$

and

$$M_k = \sup\{f(x) : \ x \in [x_{k-1}, x_k]\}.$$

The lower sum of f with respect to P is given by

$$U(f;P) = \sum_{k=1}^{n} m_k(x_k - x_{k-1}).$$

Similarly, we define the *upper sum* of f with respect to P by

$$L(f;P) = \sum_{k=1}^{n} M_k(x_k - x_{k-1}).$$

Since f is a bounded function on $[a,b]$, there exist numbers M and m such that $m \leq f(x) \leq M$ is true for all $x \in [a,b]$. Thus for any partition P of $[a,b]$ we have

$$m(b-a) \leq L(f;P) \leq U(f;P) \leq M(b-a).$$

- *Upper and Lower Integrals*: Let $\mathcal{P}$ be the collection of all possible partitions of the interval $[a,b]$. The *lower integral* of f

$$\underline{\int_a^b} f(x)\,dx = L(f) = \sup\{L(f;P) : P \in \mathcal{P}\}.$$

  Likewise the *upper integral* of f

$$\overline{\int_a^b} f(x)\,dx = U(f) = \inf\{U(f;P) : P \in \mathcal{P}\}.$$

  Clearly for a bounded function f on $[a,b]$ we always have $U(f) \geq L(f)$.

- *Riemann Integral*: A bounded function f on $[a,b]$ is *Riemann integrable* if $U(f) = L(f)$. In this case, we define $\int_a^b f(x)\,dx$ to be

$$\int_a^b f(x)\,dx = U(f) = L(f).$$

- *Riemann Sum*: Let $P = \{a = x_0 < x_1 < x_2 < \ldots < x_n = b\}$ be a partition of $[a,b]$. A *tagged partition* is one where in addition to P we have chosen points x_k^* in each of the subintervals $[x_{k-1}, x_k]$. Suppose $f : [a,b] \to \mathbb{R}$ and a tagged partition (P, x_k^*) is given. The *Riemann sum* generated by this partition is given as

$$R(f;P) = \sum_{k=1}^n f(x_k^*)(x_k - x_{k-1}).$$

  It is clear that

$$L(f;P) \leq R(f;P) \leq U(f;P)$$

  true for any bounded function.

- *Improper Integral*: Let f be defined on $[a,\infty)$ and integrable on $[a,c]$ for every $c > a$. If $\lim_{c \to \infty} \int_a^c f(x)\,dx$ exists, then the *improper integral* of f on $[a,\infty)$, denoted by $\int_a^\infty f(x)\,dx$, is given by

$$\int_a^\infty f(x)\,dx = \lim_{c \to \infty} \int_a^c f(x)\,dx.$$

- *Mean Value Theorem for Integrals*: If f is a continuous function on $[a,b]$, then there is a point $c \in (a,b)$ such that

$$\frac{1}{b-a} \int_a^b f(x)\,dx = f(c).$$

- *Fundamental Theorem of Calculus*:

a) Let $f : [a,b] \to \mathbb{R}$ be integrable, and define $F(x) = \int_a^x f$ for all $x \in [a,b]$. Then

(a) F is continuous on $[a,b]$.

(b) If f is continuous at some point $x_0 \in [a,b]$, then F is differentiable at x_0 and
$$F'(x_0) = f(x_0).$$

The above theorem expresses the fact that every continuous function is the derivative of its indefinite integral.

b) If $g : [a,b] \to \mathbb{R}$ is integrable, and $G : [a,b] \to \mathbb{R}$ satisfies $G'(x) = g(x)$ for all $x \in [a,b]$, then
$$\int_a^b g = G(b) - G(a).$$

Problem 7.1 Is the function
$$f(x) = \begin{cases} 1 & \text{if } x \neq 1, \\ 0 & \text{if } x = 1 \end{cases}$$
integrable over the interval $[0,2]$?

Problem 7.2 Let $0 < a < b$ and
$$f(x) = \begin{cases} 1 & \text{if } x \in [a,b] \cap \mathbb{Q}, \\ 0 & \text{if } x \in [a,b] \text{ is irrational.} \end{cases}$$
Find the upper and lower Riemann integrals of $f(x)$ over $[a,b]$.

Problem 7.3 Let $0 < a < b$ and
$$f(x) = \begin{cases} 0 & \text{if } x \in [a,b] \cap \mathbb{Q}, \\ x & \text{if } x \in [a,b] \text{ is irrational.} \end{cases}$$
Find the upper and lower Riemann integrals of $f(x)$ over $[a,b]$, and conclude whether $f(x)$ is Riemann integrable.

Problem 7.4 Let $0 < a < b$ and
$$f(x) = \begin{cases} \dfrac{1}{q} & \text{if } x \in [a,b] \cap \mathbb{Q} \text{ and } x = \dfrac{p}{q} \text{ where } p \text{ and } q \text{ are coprime;} \\ 0 & \text{if } x \in [a,b] \text{ is irrational.} \end{cases}$$
Show that $f(x)$ is Riemann integrable over $[a,b]$ and compute $\int_a^b f(x)dx$.

Problem 7.5 Let $f : [a,b] \to \mathbb{R}$ be a Riemann integrable function. Let $g : [a,b] \to \mathbb{R}$ be a function such that $\{x \in [a,b]; f(x) \neq g(x)\}$ is finite. Show that $g(x)$ is Riemann integrable and
$$\int_a^b f(x)dx = \int_a^b g(x)dx \ .$$
Does the conclusion still hold when $\{x \in [a,b]; f(x) \neq g(x)\}$ is countable?

Problem 7.6 Let $f : [a,b] \to \mathbb{R}$ be a Riemann integrable function. Show that $|f(x)|$ is Riemann integrable and
$$\left| \int_a^b f(x)dx \right| \leq \int_a^b |f(x)|dx \ .$$
When do we have equality?

Problem 7.7 We call $U : [a,b] \to \mathbb{R}$ a step function if there is a partition $\mathcal{P}$ of $[a,b]$ so that U is constant on each interval of $\mathcal{P}$. Show that any function $f : [a,b] \to \mathbb{R}$ is Riemann integrable if and only if for each $\varepsilon > 0$, there exist two step functions U and V on $[a,b]$ such that $V(x) \leq f(x) \leq U(x)$ and
$$\int_a^b \Big(U(x) - V(x)\Big)dx < \varepsilon.$$

Problem 7.8 (**Second Mean Value Theorem for Integrals**) Prove that if f and g are defined on $[a,b]$ with g continuous, $f \geq 0$, and f is integrable, then there is a point $x_0 \in (a,b)$ such that
$$\int_a^b f(x)g(x)dx = g(x_0) \int_a^b f(x)dx.$$

Problem 7.9 Use Riemann sums to find the following limits:

(i) $\displaystyle\lim_{n \to \infty} \sum_{k=1}^{k=n} \frac{n}{4n^2 + k^2}$

(ii) $\displaystyle\lim_{n \to \infty} \sum_{k=1}^{k=n} \arctan\left(\frac{n}{n^2 + k^2}\right)$

  Hint: use the inequalities $x - \dfrac{x^3}{3} < \arctan(x) < x$, for $x > 0$.

(iii) $\displaystyle\lim_{n \to \infty} \sum_{k=1}^{k=n} \frac{1}{k} \tan\left(\frac{k\pi}{4n+4}\right)$

  Hint: use the function $f(x) = \dfrac{\tan(x)}{x}$.

CHAPTER 7. INTEGRATION

Problem 7.10 Let $f : [0,1] \to \mathbb{R}$ be C^1, i.e., $f(x)$ is differentiable and its derivative is continuous. Find
$$\lim_{n \to \infty} \sum_{k=1}^{k=n} f\left(\frac{k}{n}\right) - n \int_0^1 f(x)dx \ .$$
Hint: use $\int_0^1 f(x)dx = \sum_{k=1}^{k=n} \int_{(k-1)/n}^{k/n} f(x)dx$.

Problem 7.11 Let $f : [0,1] \to \mathbb{R}$ be C^2, i.e., $f(x)$ is twice differentiable and $f''(x)$ is continuous. Find
$$\lim_{n \to \infty} n^2 \int_0^1 f(x)dx - n \sum_{k=1}^{k=n} f\left(\frac{2k-1}{2n}\right) \ .$$

Problem 7.12 Define
$$x_n = \sum_{k=1}^{n} \frac{2}{2n+2k-1} \ .$$
Find $\lim_{n \to \infty} x_n$ and $\lim_{n \to \infty} n^2(\ln(2) - x_n)$.

Problem 7.13 Find the following limits:

(i) $\lim_{n \to \infty} \int_0^1 \frac{x^n}{1+x}dx$

(ii) $\lim_{n \to \infty} \int_0^1 nx(1-x^2)^n dx$

(iii) $\lim_{n \to \infty} \int_1^2 \frac{\sin(nx)}{x}dx$

Problem 7.14 Show that
$$\int_0^\pi xe^{\sin(x)}dx = \frac{\pi}{2} \int_0^\pi e^{\sin(x)}dx \ .$$

Problem 7.15 Let $f : [a,b] \to \mathbb{R}$ be a continuous function. Show that there exists $c \in (a,b)$ such that
$$\frac{1}{b-a} \int_a^b f(x)dx = f(c) \ .$$
Is this still true for Riemann integrable functions?

Problem 7.16 Let $f : \mathbb{R} \to \mathbb{R}$ be continuous and set $F(x) = \int_0^{x^3} f(y)\,dy$. Show that
$$F'(x) = 3x^2 f(x^3).$$

Problem 7.17 Consider the function
$$f(x) = \begin{cases} 1 & 0 \leq x \leq 1, \\ 0 & 1 < x \leq 2. \end{cases}$$

(a) What is $F(x) = \int_0^x f(t)\,dt$ on $[0,2]$?

(b) Is $F(x)$ continuous?

(c) Is $F'(x) = f(x)$?

Problem 7.18 Let $f : [a,b] \to \mathbb{R}$ be a continuous function. Show that
$$\lim_{n \to \infty} \int_a^b f(x) \sin(nx)\,dx = 0 \text{ and } \lim_{n \to \infty} \int_a^b f(x) \cos(nx)\,dx = 0.$$
Use these limits to find
$$\lim_{n \to \infty} \int_a^b f(x) \sin^2(nx)\,dx.$$
This is known as Riemann–Lebesgue's lemma.

Problem 7.19 Let $f : [a,b] \to \mathbb{R}$ be a Riemann integrable function. Show that if $\int_a^b f^2(x)\,dx = 0$, then $f(x_0) = 0$ whenever $f(x)$ is continuous at $x_0 \in (a,b)$. What can we say about the set
$$\{x \in [a,b]; f(x) \neq 0\}?$$

Problem 7.20 Decide whether or not the following integrals converge or diverge:

(a) $\int_1^\infty \dfrac{1}{\sqrt{1+x^3}}\,dx$.

(b) $\int_1^\infty \dfrac{\sin(x)}{x}\,dx$. Does this integral converge absolutely?

(c) $\int_0^\infty \dfrac{x}{1+x^2\sin^2(x)}\,dx$.

CHAPTER 7. INTEGRATION

Problem 7.21 (**Bertrand Integrals**) Discuss the convergence or divergence of the Bertrand Integrals
$$\int_2^\infty \frac{1}{x^\alpha \ln^\beta(x)} dx$$
depending on the parameters α and β.

Problem 7.22 (**Cauchy Criterion**) Let $f : [a, \infty) \to \mathbb{R}$ be Riemann integrable on bounded intervals. Show that $\int_a^\infty f(x)dx$ converges if and only if for every $\varepsilon > 0$ there exists $A > a$ such that for any $t_1, t_2 > A$ we have
$$\left| \int_{t_1}^{t_2} f(x)dx \right| < \varepsilon .$$

Problem 7.23 (**Abel's test**) Assume that the functions f and g defined on $[a, \infty)$ satisfy the following conditions:

(a) g is monotone and bounded on $[a, \infty)$,

(b) the improper integral $\int_a^\infty f(x)dx$ is convergent.

Then prove that $\int_a^\infty f(x)g(x)dx$ is also convergent.

Problem 7.24 (**Luxemburg Monotone Convergence Theorem**) Let $\{f_n(x)\}$ be a decreasing sequence of bounded functions on $[a, b]$, with $a < b$, which converges pointwise to 0 on $[a, b]$. Show that
$$\lim_{n \to \infty} \int_a^b f_n(x)dx = 0.$$

Problem 7.25 (**Monotone Convergence Theorem**) Let $\{f_n(x)\}$ be a decreasing sequence of Riemann integrable functions on $[a, b]$ which converges pointwise to a Riemann integrable function $f(x)$. Show that
$$\lim_{n \to \infty} \int_a^b f_n(x)dx = \int_a^b f(x)dx.$$

Problem 7.26 (**Arzelà Theorem**) Let $\{f_n(x)\}$ be a sequence of Riemann integrable functions on $[a,b]$ which converges pointwise to a Riemann integrable function $f(x)$. Assume that there exists $M > 0$ such that $|f_n(x)| \leq M$ for any $x \in [a,b]$ and $n \geq 1$. Show that

$$\lim_{n \to \infty} \int_a^b f_n(x) dx = \int_a^b f(x) dx.$$

Problem 7.27 (**Fatoo Lemma**) Let $\{f_n(x)\}$ be a sequence of Riemann integrable functions on $[a,b]$ which converges pointwise to a Riemann integrable function $f(x)$. Show that

$$\int_a^b f(x) dx \leq \liminf_{n \to \infty} \int_a^b f_n(x) dx.$$

Problem 7.28 Let $T : (C[0,1], \| \cdot \|) \to (C[0,1], \| \cdot \|)$ be a function defined as

$$Tf(x) = \int_0^x f(t) dt$$

where by $C[0,1]$ we mean the vector space of all continuous real-valued functions defined on $[0,1]$ and $\| f \| = \sup_{0 \leq t \leq 1} |f(t)|$. Show that:

a) T is **not** a contraction.

b) T has a unique fixed point.

c) T^2 is a contraction.

Problem 7.29 Convert the initial value problem

$$\frac{dy}{dx} = 3xy, \quad y(0) = 1$$

to an integral equation. Use Picard's iteration scheme to solve it.
Note: For the initial value problem $\frac{dy}{dx} = f(x,y)$, $y(x_0) = y_0$, Picard iteration defined by

$$y_{n+1}(x) = y_0 + \int_{x_0}^x f(t, y_n(t)) \, dt \text{ for } n \in \mathbb{Z}^+.$$

Problem 7.30 Given the initial value problem

$$f'(x) = 1 + x - f(x), \text{ with } f(0) = 1,$$

for $x \in \left[-\frac{1}{2}, \frac{1}{2}\right]$. Show the mapping $T : C[-\frac{1}{2}, \frac{1}{2}] \to C[-\frac{1}{2}, \frac{1}{2}]$ defined by

$$Tf(x) = 1 + x + \frac{1}{2}x^2 - \int_0^x f(t)dt$$

is a contraction. Then set up Picard's iteration scheme to solve it.

Problem 7.31 Show that the integral $\int_0^\pi \frac{\sin xt}{t} dt$ depends continuously on x.

Problem 7.32 (Tchebycheff polynomials)

1. For any $n \in \mathbb{N}$, find a polynomial T_n such that $T_n(\cos(x)) = \cos(nx)$. Find the degree of T_n, $n \in \mathbb{N}$.

2. Show that for any $n \geq 1$, we have

$$T_{n+1} = 2xT_n - T_{n-1}.$$

3. Find the integrals

$$\int_{-1}^1 \frac{T_n(x)T_m(x)}{\sqrt{1-x^2}} dx,$$

for any $n, m \in \mathbb{N}$.

4. Show that $\cos\left(\frac{(2k-1)\pi}{2n}\right)$, $k \in [1, n]$, are n different roots of $T_n(x)$, for any $n \geq 1$.

Solutions

Solution 7.1

If $\mathcal{P}$ is any partition of $[0,2]$, then

$$U(\mathcal{P},f) = \sum_{k=1}^{n} M_k(x_k - x_{k-1}) = 2.$$

$L(f,\mathcal{P})$ will be less than 2, because any subinterval of $\mathcal{P}$ that contains $x = 1$ will contribute zero to the value of the lower sum. The way to show that f is integrable is to construct a partition that minimizes the effect of the discontinuity by embedding $x = 1$ into a very small subinterval. Let $\varepsilon > 0$, and consider the partition $\mathcal{P}_\varepsilon = \{0,\ 1 - \varepsilon/3,\ 1 + \varepsilon/3,\ 2\}$. Then

$$\begin{aligned}L(\mathcal{P}_\varepsilon, f) &= 1[(1 - \varepsilon/3) - 0] + 0[1 + \varepsilon/3 - (1 - \varepsilon/3)] + 1[2 - (1 + \varepsilon/3)] \\ &= 1(1 - \varepsilon/3) + 0(2\varepsilon/3) + 1(1 - \varepsilon/3) \\ &= 2 - 2\varepsilon/3.\end{aligned}$$

Now, $U(\mathcal{P}_\varepsilon, f) = 2$, so we have

$$U(\mathcal{P}_\varepsilon, f) - L(\mathcal{P}_\varepsilon, f) = 2\varepsilon/3 < \varepsilon.$$

Thus, f is integrable.

Solution 7.2

Let $\mathcal{P}$ be a partition of $[a,b]$. If $\mathcal{P} = \{x_1 = a < x_2 < \ldots < x_n = b\}$, then

$$\inf\{f(x); x_i < x < x_{i+1}\} = 0 \quad \text{and} \quad \sup\{f(x); x_i < x < x_{i+1}\} = 1$$

because $\mathbb{Q}$ and $\mathbb{R} \setminus \mathbb{Q}$ are dense in $\mathbb{R}$. Hence we have

$$L(\mathcal{P}, f) = 0 \quad \text{and} \quad U(\mathcal{P}, f) = b - a\ .$$

Obviously this will imply

$$\underline{\int_a^b} f(x)dx = \sup L(\mathcal{P}, f) = 0$$

and

$$\overline{\int_a^b} f(x)dx = \inf U(\mathcal{P}, f) = b - a\ ,$$

where the infimum and supremum are taken over all partitions of $[a,b]$. So clearly $f(x)$ is not Riemann integrable on $[a,b]$.

Solution 7.3

Let $\mathcal{P}$ be a partition of $[a,b]$. If $\mathcal{P} = \{x_1 = a < x_2 < \ldots < x_n = b\}$, then

$$\inf\{f(x); x_i < x < x_{i+1}\} = 0 \quad \text{and} \quad \sup\{f(x); x_i < x < x_{i+1}\} = x_{i+1}$$

CHAPTER 7. INTEGRATION

because $\mathbb{Q}$ and $\mathbb{R} \setminus \mathbb{Q}$ are dense in $\mathbb{R}$. Hence we have $L(\mathcal{P}, f) = 0$, and

$$U(\mathcal{P}, f) = x_2(x_2 - x_1) + \cdots + x_i(x_i - x_{i-1}) + \cdots + x_n(x_n - x_{n-1}) = U(\mathcal{P}, h),$$

where $h : [a, b] \to \mathbb{R}$ defined by $h(x) = x$. Since $h(x)$ is continuous, $h(x)$ is Riemann integrable and

$$\overline{\int_a^b} h(x) dx = \inf U(\mathcal{P}, h) = \frac{b^2 - a^2}{2},$$

where the infimum is taken over all partitions of $[a, b]$. Hence

$$\underline{\int_a^b} f(x) dx = \sup L(\mathcal{P}, f) = 0$$

and

$$\overline{\int_a^b} f(x) dx = \inf U(\mathcal{P}, f) = \frac{b^2 - a^2}{2}.$$

Since $0 < \frac{b^2 - a^2}{2}$, we conclude that $f(x)$ is not Riemann integrable on $[a, b]$.

$\boxed{\text{Solution 7.4}}$

Let $\mathcal{P}$ be a partition of $[a, b]$. If $\mathcal{P} = \{x_1 = a < x_2 < \ldots < x_n = b\}$, then

$$\inf\{f(x); x_i < x < x_{i+1}\} = 0$$

because $\mathbb{R} \setminus \mathbb{Q}$ is dense in $\mathbb{R}$. Hence we have $L(\mathcal{P}, f) = 0$, which implies

$$\underline{\int_a^b} f(x) dx = \sup L(\mathcal{P}, f) = 0,$$

where the supremum is taken over all partitions of $[a, b]$. Let us now show that

$$\overline{\int_a^b} f(x) dx = \inf U(\mathcal{P}, f) = 0$$

where the infimum is taken over all partitions of $[a, b]$. Fix $\varepsilon > 0$. Then the set

$$B_\varepsilon = \left\{ \frac{p}{q} \in [a, b] \cap \mathbb{Q}; \text{ where } p \text{ and } q \text{ are coprime and } \frac{1}{q} \geq \frac{\varepsilon}{2(b - a)} \right\}$$

is finite. Without loss of generality, assume that B_ε is not empty and has $n \geq 1$ elements. Set

$$B_\varepsilon = \{x_1 < x_2 < \ldots < x_n\}.$$

Assume for now $a < x_1$ and $x_n < b$. Choose $m \geq 1$ large enough to have $\frac{1}{m} < \frac{\varepsilon}{2n}$, $x_i + \frac{1}{2m} < x_{i+1} - \frac{1}{2m}$, $a < x_1 - \frac{1}{2m}$, and $x_n + \frac{1}{2m} < b$. Consider the partition

$$\mathcal{P}_0 = \left\{ a, x_i - \frac{1}{2m}, x_i + \frac{1}{2m}, b \right\}.$$

We have
$$0 \leq \sup\left\{f(x);\ x_i - \frac{1}{2m} \leq x \leq x_i + \frac{1}{2m}\right\} \leq 1$$
because $0 \leq f(x) \leq 1$ for all $x \in [a,b]$. On any other interval I associated with the partition, we have
$$0 \leq \sup\{f(x);\ x \in I\} \leq \frac{\varepsilon}{2(b-a)}$$
because $I \cap B_\varepsilon$ is empty. Hence
$$0 \leq U(\mathcal{P}_0, f) \leq n\frac{1}{m} + \frac{\varepsilon}{2(b-a)}(b-a) = \frac{\varepsilon}{2} + \frac{\varepsilon}{2} = \varepsilon\ .$$

If $x_1 = a$, then we consider only the interval $[a, a+1/m]$ and if $x_n = b$, then we consider only the interval $[b-1/m, b]$. The proof is carried similarly to get $U(\mathcal{P}_0, f) \leq \varepsilon$. Clearly this will imply $\inf U(\mathcal{P}, f) = 0$, where the infimum is taken over all partitions of $[a,b]$. Hence
$$\underline{\int_a^b} f(x)dx = \overline{\int_a^b} f(x)dx = 0\ ,$$
which implies that $f(x)$ is Riemann integrable and $\int_a^b f(x)dx = 0$.

Solution 7.5

Consider the function $h(x) = g(x) - f(x)$. Fix $\varepsilon > 0$. Then the set $B = \{x \in [a,b]; h(x) \neq 0\}$ is finite. Assume that $B = \{x_1, \ldots, x_n\}$ with $a < x_1 < \ldots < x_n < b$. Set
$$M = \max\{|h(x_i)| : i = 1, 2, \ldots, n\}\ .$$

We have $M > 0$. Choose $\delta > 0$ small enough to have $a < x_1 - \delta$, $x_i + \delta < x_{i+1} - \delta$, $x_n + \delta < b$, and $\delta < \dfrac{\varepsilon}{2Mn}$. Consider the partition
$$\mathcal{P}_0 = \{a, x_i - \delta, x_i + \delta, b\}\ .$$

We have
$$0 \leq \sup\{|h(x)| :\ x_i - \delta \leq x \leq x_i + \delta\} \leq M$$
because $0 \leq |h(x)| \leq M$ for all $x \in [a,b]$. On any other interval I associated with the partition, we have
$$\sup\{|h(x)| :\ x \in I\} = 0$$
because $I \cap B$ is empty. Hence
$$0 \leq |U(\mathcal{P}_0, h)| \leq nM2\delta \leq \varepsilon\ ,\text{ and }\ 0 \leq |L(\mathcal{P}_0, h)| \leq nM2\delta \leq \varepsilon\ .$$

If $x_1 = a$, then we consider only the interval $[a, a+\delta]$ and if $x_n = b$, then we consider only the interval $[b-\delta, b]$. The proof is carried similarly to get $|U(\mathcal{P}_0, h)| \leq \varepsilon$ and $|L(\mathcal{P}_0, h)| \leq \varepsilon$. Clearly this will imply $\inf U(\mathcal{P}, h) = 0$ and $\sup L(\mathcal{P}, h) = 0$, where the infimum and supremum are taken over all partitions of $[a,b]$. Hence
$$\underline{\int_a^b} h(x)dx = \overline{\int_a^b} h(x)dx = 0\ ,$$

CHAPTER 7. INTEGRATION

which implies that $h(x)$ is Riemann integrable and $\int_a^b h(x)dx = 0$. Therefore $g(x) = f(x) + h(x)$ is also Riemann integrable and

$$\int_a^b g(x)dx = \int_a^b f(x)dx + \int_a^b h(x)dx = \int_a^b f(x)dx \ .$$

Finally, note that if $\{x \in [a,b]; f(x) \neq g(x)\}$ is countable but not finite, then the conclusion may not be true. Indeed, take

$$f(x) = \begin{cases} 1 & \text{if } x \in [a,b] \cap \mathbb{Q}, \\ 0 & \text{if } x \in [a,b] \text{ is irrational,} \end{cases}$$

and $g(x) = 0$. Then $\{x \in [a,b]; f(x) \neq g(x)\} = [a,b] \cap \mathbb{Q}$ which is countable. But $f(x)$ is not Riemann integrable while $g(x)$ is.

Solution 7.6

First, let us note that if $f : [a,b] \to \mathbb{R}$ is Riemann integrable and $h : [c,d] \to \mathbb{R}$ is continuous, then $h \circ f(x) = h(f(x))$ is Riemann integrable provided $f([a,b]) \subset [c,d]$. Since $f(x)$ is Riemann integrable, it is bounded. Hence there exist $c, d \in \mathbb{R}$ such that $f([a,b]) \subset [c,d]$. Set $h(x) = |x|$. Since $h(x)$ is continuous on any closed interval, $|f(x)|$ is Riemann integrable. Another way to see this is true is by the characterization of Riemann integrability. Indeed, we know that a function is Riemann integrable if and only if it is continuous except maybe at a set which is negligible or has measure 0. So if we assume that $f(x) : [a,b] \to \mathbb{R}$ is Riemann integrable, then the set $\{x_0 \in [a,b] : f(x) \text{ is not continuous at } x_0\}$ has measure 0. But it is known that if $f(x)$ is continuous at $x_0 \in [a,b]$, then $|f(x)|$ is also continuous at x_0. This clearly implies that the set $\{x_0 \in [a,b]; |f(x)| \text{ is not continuous at } x_0\}$ has measure 0 as well because a subset of a measure 0 set has measure 0. Hence $|f(x)|$ is Riemann integrable. Once this fact is established, we can prove the inequality

$$\left| \int_a^b f(x)dx \right| \leq \int_a^b |f(x)|dx \ .$$

One way is to use the inequalities $-|f(x)| \leq f(x) \leq |f(x)|$ to get

$$\int_a^b -|f(x)|dx \leq \int_a^b f(x)dx \leq \int_a^b |f(x)|dx \ ,$$

which will lead to the desired conclusion. Let us give another proof of this inequality. Indeed, consider the partition

$$\mathcal{P} = \left\{ a_i; \ a_i = a + i\frac{(b-a)}{n} \text{ for } i = 1, 2, \ldots, n \right\} \ .$$

Then we have

$$\lim_{n \to \infty} \sum_{i=1}^{i=n} \frac{(b-a)}{n} f(a_i) = \int_a^b f(x)dx$$

and

$$\lim_{n \to \infty} \sum_{i=1}^{i=n} \frac{(b-a)}{n} |f(a_i)| = \int_a^b |f(x)|dx \ .$$

Since
$$\left|\sum_{i=1}^{i=n}\frac{(b-a)}{n}f(a_i)\right| \leq \sum_{i=1}^{i=n}\frac{(b-a)}{n}|f(a_i)|,$$
we get
$$\left|\int_a^b f(x)dx\right| \leq \int_a^b |f(x)|dx.$$

Solution 7.7

Indeed assume that $f(x)$ is Riemann integrable. Then for any $\varepsilon > 0$, there exists a partition $\mathcal{P}$ such that $U(\mathcal{P},f) - L(\mathcal{P},f) < \varepsilon$. Define the two step functions $U(x)$ and $V(x)$ by
$$U(x) = \sup\{f(z); z \in I\} \text{ and } V(x) = \inf\{f(z); z \in I\},$$
for $x \in I$, where I is any interval of the partition $\mathcal{P}$. Then we have $V(x) \leq f(x) \leq U(x)$, for any $x \in [a,b]$, by definition of these two step functions. And since
$$\int_a^b U(x)dx = U(\mathcal{P},f) \text{ and } \int_a^b V(x)dx = L(\mathcal{P},f),$$
we get
$$\int_a^b \Big(U(x) - V(x)\Big)dx < \varepsilon.$$
Now assume the converse is true, i.e., for each $\varepsilon > 0$, there exist two step functions U and V on $[a,b]$ such that $V(x) \leq f(x) \leq U(x)$ and
$$\int_a^b \Big(U(x) - V(x)\Big)dx < \varepsilon.$$
Let us prove that $f(x)$ is Riemann integrable. Since $V(x) \leq f(x) \leq U(x)$, we easily get
$$U(\mathcal{P},f) \leq \int_a^b U(x)dx \text{ and } \int_a^b V(x)dx \leq L(\mathcal{P},f).$$
Hence $U(\mathcal{P},f) - L(\mathcal{P},f) < \varepsilon$, which implies the desired conclusion.

Solution 7.8

Since g is continuous on the compact interval $[a,b]$, we know $m = \inf(g([a,b]))$ and $M = \sup(g([a,b]))$ exists as finite real numbers and that there are points x_1, x_2 in $[a,b]$ such that $g(x_1) = m$ and $g(x_2) = M$. Since
$$m \leq g(x) \leq M \text{ and } f(x) \geq 0,$$
we have
$$mf(x) \leq f(x)g(x) \leq Mf(x) \text{ for all } x \in [a,b].$$
Then assuming f and $f \cdot g$ are integrable on $[a,b]$ we have
$$m\int_a^b f(x)dx \leq \int_a^b f(x)g(x)dx \leq M\int_a^b f(x)dx.$$

Next observe that the function $h(t) = t\int_a^b f(x)dx$ depends continuously on t where $t \in [m, M]$, furthermore $\int_a^b f(x)g(x)dx$ is in $[h(m), h(M)]$. By the intermediate value theorem there is a number $t_0 \in [m, M]$ with $h(t_0) = \int_a^b f(x)g(x)dx$. Since g is continuous between x_1 and x_2 and $t_0 \in [m, M] = [g(x_1), g(x_2)]$ a second application of the intermediate value theorem to g this time yields $g(x_0) = t_0$, therefore

$$\int_a^b f(x)g(x)dx = h(t_0) = t_0 \int_a^b f(x)dx = g(x_0)\int_a^b f(x)dx.$$

as claimed. The assumption that the product of $f \cdot g$ is integrable on $[a, b]$ follows from the fact that if $f : [a, b] \to \mathbb{R}$ is integrable and $g : [a, b] \to \mathbb{R}$ is continuous, then the product $f \cdot g : [a, b] \to \mathbb{R}$ is also integrable on $[a, b]$.

$\boxed{\text{Solution 7.9}}$

Let us find the limit at (i). Consider the function $f(x) = \dfrac{1}{4+x^2}$. Then we have

$$\lim_{n\to\infty} \sum_{k=1}^{k=n} \frac{1}{n} f\left(\frac{k}{n}\right) = \int_0^1 f(x)dx \ .$$

But

$$\sum_{k=1}^{k=n} \frac{1}{n} f\left(\frac{k}{n}\right) = \sum_{k=1}^{k=n} \frac{1}{n} \frac{1}{4+\left(\frac{k}{n}\right)^2} = \sum_{k=1}^{k=n} \frac{n}{4n^2+k^2} \ ,$$

which implies

$$\lim_{n\to\infty} \sum_{k=1}^{k=n} \frac{n}{4n^2+k^2} = \int_0^1 f(x)dx = \frac{1}{2}\arctan\left(\frac{1}{2}\right) \ .$$

Next let us find the limit at (ii). Using the hint we get

$$\frac{n}{n^2+k^2} - \frac{1}{3}\left(\frac{n}{n^2+k^2}\right)^3 \leq \arctan\left(\frac{n}{n^2+k^2}\right) \leq \frac{n}{n^2+k^2} \ .$$

As for the previous limit (in (i)), one will easily show

$$\lim_{n\to\infty} \sum_{k=1}^{k=n} \frac{n}{n^2+k^2} = \int_0^1 \frac{1}{1+x^2}dx = \frac{\pi}{4} \ .$$

On the other hand, we have

$$\sum_{k=1}^{k=n} \left(\frac{n}{n^2+k^2}\right)^3 \leq \sum_{k=1}^{k=n} \left(\frac{n}{n^2}\right)^3 = \sum_{k=1}^{k=n} \frac{1}{n^3} = \frac{1}{n^2} \ ,$$

which implies

$$\lim_{n\to\infty} \sum_{k=1}^{k=n} \left(\frac{n}{n^2+k^2}\right)^3 = 0 \ .$$

The Squeeze Theorem will then imply

$$\lim_{n\to\infty} \sum_{k=1}^{k=n} \arctan\left(\frac{n}{n^2+k^2}\right) = \frac{\pi}{4}.$$

For the last limit, i.e., in (iii), consider the function $f(x) = \dfrac{\tan(x)}{x}$. Then we have

$$\sum_{k=1}^{k=n} \frac{1}{k} \tan\left(\frac{k\pi}{4n+4}\right) = \sum_{k=1}^{k=n} \frac{\pi}{4n+4} f\left(\frac{k\pi}{4n+4}\right).$$

Using the Riemann sums, we get

$$\lim_{n\to\infty} \sum_{k=1}^{k=n} \frac{\pi}{4n+4} f\left(\frac{k\pi}{4n+4}\right) = \int_0^{\frac{\pi}{4}} f(x)dx,$$

or

$$\lim_{n\to\infty} \sum_{k=1}^{k=n} \frac{1}{k} \tan\left(\frac{k\pi}{4n+4}\right) = \int_0^{\frac{\pi}{4}} \frac{\tan(x)}{x} dx.$$

Note that the function $f(x)$, extended at $x=0$ by setting $f(0)=1$, is continuous on $[0,a]$, where $a > 0$ is any number. Hence $\int_0^{\pi/4} \dfrac{\tan(x)}{x} dx$ exists in $\mathbb{R}$.

Solution 7.10

Using the hint, we get

$$\sum_{k=1}^{k=n} f\left(\frac{k}{n}\right) - n\int_0^1 f(x)dx = n\sum_{k=1}^{k=n} \frac{1}{n} f\left(\frac{k}{n}\right) - n\sum_{k=1}^{k=n} \int_{(k-1)/n}^{k/n} f(x)dx,$$

which yields

$$\sum_{k=1}^{k=n} f\left(\frac{k}{n}\right) - n\int_0^1 f(x)dx = n\sum_{k=1}^{k=n} \int_{(k-1)/n}^{k/n} \left(f\left(\frac{k}{n}\right) - f(x)\right) dx.$$

Since $f(x)$ is C^1, the Mean Value Theorem implies

$$\int_{(k-1)/n}^{k/n} \left(f\left(\frac{k}{n}\right) - f(x)\right) dx = \int_{(k-1)/n}^{k/n} f'(\theta_k(x)) \left(\frac{k}{n} - x\right) dx,$$

for any $k = 1, 2, \ldots, n$. Because $f'(x)$ is continuous on $[0,1]$, it is bounded. Set

$$m_k = \inf\left\{f'(x); \frac{(k-1)}{n} \leq x \leq \frac{k}{n}\right\} \text{ and } M_k = \sup\left\{f'(x); \frac{(k-1)}{n} \leq x \leq \frac{k}{n}\right\}.$$

Hence we have

$$\int_{(k-1)/n}^{k/n} m_k \left(\frac{k}{n} - x\right) dx \leq \int_{(k-1)/n}^{k/n} \left(f\left(\frac{k}{n}\right) - f(x)\right) dx \leq \int_{(k-1)/n}^{k/n} M_k \left(\frac{k}{n} - x\right) dx,$$

CHAPTER 7. INTEGRATION

for all $k = 1, 2, \ldots, n$. Since $\displaystyle\int_{(k-1)/n}^{k/n} \left(\frac{k}{n} - x\right) dx = \frac{1}{2n^2}$, we get

$$\sum_{k=1}^{k=n} \frac{m_k}{2n^2} \leq \sum_{k=1}^{k=n} \int_{(k-1)/n}^{k/n} \left(f\left(\frac{k}{n}\right) - f(x)\right) dx \leq \sum_{k=1}^{k=n} \frac{M_k}{2n^2}.$$

So

$$\sum_{k=1}^{k=n} \frac{m_k}{2n} \leq \sum_{k=1}^{k=n} f\left(\frac{k}{n}\right) - n \int_0^1 f(x)dx \leq \sum_{k=1}^{k=n} \frac{M_k}{2n}.$$

Since

$$\sum_{k=1}^{k=n} \frac{m_k}{n} \quad \text{and} \quad \sum_{k=1}^{k=n} \frac{M_k}{n}$$

are the upper and lower Riemann sums associated to the function $f'(x)$ and the partition $\mathcal{P} = \{k/n; \ k = 1, 2, \ldots, n\}$ of $[0, 1]$, we get

$$\lim_{n \to \infty} \sum_{k=1}^{k=n} \frac{m_k}{n} = \int_0^1 f'(x)dx = f(1) - f(0), \text{ and } \lim_{n \to \infty} \sum_{k=1}^{k=n} \frac{M_k}{n} = \int_0^1 f'(x)dx = f(1) - f(0).$$

The Squeeze Theorem will then force the equality

$$\lim_{n \to \infty} \sum_{k=1}^{k=n} f\left(\frac{k}{n}\right) - n \int_0^1 f(x)dx = \frac{f(1) - f(0)}{2}.$$

Solution 7.11

The proof will follow the same ideas as the ones developed in the previous problem. Indeed, we know from Taylor's formula that

$$f(y) = f(x) + f'(x)(y - x) + \frac{f''(\theta)}{2}(y - x)^2,$$

for any $x, y \in [0, 1]$ for some θ between x and y. If we use the same hint as in the previous problem, we get

$$n^2 \int_0^1 f(x)dx - n \sum_{k=1}^{k=n} f\left(\frac{2k-1}{2n}\right) = n^2 \left(\sum_{k=1}^{k=n} \int_{(k-1)/n}^{k/n} f(x)dx - \frac{1}{n} f\left(\frac{2k-1}{2n}\right)\right),$$

or

$$n^2 \int_0^1 f(x)dx - n \sum_{k=1}^{k=n} f\left(\frac{2k-1}{2n}\right) = n^2 \left(\sum_{k=1}^{k=n} \int_{(k-1)/n}^{k/n} \left(f(x) - f\left(\frac{2k-1}{2n}\right)\right) dx\right).$$

Using Taylor's formula, we get

$$f(x) - f\left(\frac{2k-1}{2n}\right) = f'\left(\frac{2k-1}{2n}\right)\left(x - \frac{2k-1}{2n}\right) + \frac{f''(\theta_k(x))}{2}\left(x - \frac{2k-1}{2n}\right)^2.$$

Since

$$\int_{(k-1)/n}^{k/n} f'\left(\frac{2k-1}{2n}\right)\left(x - \frac{2k-1}{2n}\right) dx = 0,$$

we get

$$n^2 \int_0^1 f(x)dx - n\sum_{k=1}^{k=n} f\left(\frac{2k-1}{2n}\right) = n^2 \sum_{k=1}^{k=n} \int_{(k-1)/n}^{k/n} \frac{f''(\theta_k(x))}{2}\left(x - \frac{2k-1}{2n}\right)^2 dx .$$

As we did in the previous problem, set

$$m_k = \inf\left\{f''(x); \frac{(k-1)}{n} \leq x \leq \frac{k}{n}\right\} \text{ and } M_k = \sup\left\{f''(x); \frac{(k-1)}{n} \leq x \leq \frac{k}{n}\right\} .$$

Hence

$$\frac{m_k}{2}\left(x - \frac{2k-1}{2n}\right)^2 \leq \frac{f''(\theta_k(x))}{2}\left(x - \frac{2k-1}{2n}\right)^2 \leq \frac{M_k}{2}\left(x - \frac{2k-1}{2n}\right)^2 ,$$

for any $x \in [(k-1)/n, k/n]$. Since

$$\int_{(k-1)/n}^{k/n} \left(x - \frac{2k-1}{2n}\right)^2 dx = \frac{1}{12n^3} ,$$

we get

$$\sum_{k=1}^{k=n} \frac{m_k}{24n} \leq n^2 \sum_{k=1}^{k=n} \int_{(k-1)/n}^{k/n} \frac{f''(\theta_k(x))}{2}\left(x - \frac{2k-1}{2n}\right)^2 dx \leq \sum_{k=1}^{k=n} \frac{M_k}{24n} .$$

Hence

$$\sum_{k=1}^{k=n} \frac{m_k}{24n} \leq n^2 \int_0^1 f(x)dx - n\sum_{k=1}^{k=n} f\left(\frac{2k-1}{2n}\right) \leq \sum_{k=1}^{k=n} \frac{M_k}{24n} .$$

Since

$$\sum_{k=1}^{k=n} \frac{m_k}{n} \text{ and } \sum_{k=1}^{k=n} \frac{M_k}{n}$$

are the upper and lower Riemann sums associated to the function $f''(x)$ and the partition $\mathcal{P} = \{k/n;\ k = 1, 2, \ldots, n\}$ of $[0,1]$, we get

$$\lim_{n\to\infty} \sum_{k=1}^{k=n} \frac{m_k}{n} = \int_0^1 f''(x)dx = f'(1) - f'(0), \text{ and } \lim_{n\to\infty} \sum_{k=1}^{k=n} \frac{M_k}{n} = \int_0^1 f''(x)dx = f'(1) - f'(0).$$

The Squeeze Theorem will then force the equality

$$\lim_{n\to\infty} n^2 \int_0^1 f(x)dx - n\sum_{k=1}^{k=n} f\left(\frac{2k-1}{2n}\right) = \frac{f'(1) - f'(0)}{24} .$$

Solution 7.12

Set $f(x) = \dfrac{1}{x+1}$. Then

$$x_n = \frac{1}{n}\sum_{k=1}^n \frac{2n}{2n+2k-1} = \frac{1}{n}\sum_{k=1}^n f\left(\frac{2k-1}{2n}\right) .$$

Since $f(x)$ is continuous on $[0,1]$, we recognize a Riemann sum which implies

$$\lim_{n \to \infty} x_n = \int_0^1 f(x)dx = \ln(2) \ .$$

Since

$$n^2(\ln(2) - x_n) = n^2 \int_0^1 f(x)dx - n \sum_{k=1}^n f\left(\frac{2k-1}{2n}\right) \ ,$$

the previous problem implies

$$\lim_{n \to \infty} n^2(\ln(2) - x_n) = \frac{f'(1) - f'(0)}{24} = \frac{1}{32} \ .$$

Solution 7.13

For (i), intuitively since x^n goes to 0 almost everywhere on $[0,1]$, then maybe the limit of the integrals is the integral of the constant function 0, which is 0. Let us try to prove it. Fix $\delta \in (0,1)$. Then we have

$$\int_0^1 \frac{x^n}{1+x}dx = \int_0^\delta \frac{x^n}{1+x}dx + \int_\delta^1 \frac{x^n}{1+x}dx \ .$$

Since

$$0 \leq \int_0^\delta \frac{x^n}{1+x}dx \leq \int_0^\delta x^n dx \leq \delta^n \int_0^\delta dx = \delta^{n+1} \ ,$$

we obtain

$$\lim_{n \to \infty} \int_0^\delta \frac{x^n}{1+x}dx = 0 \ .$$

So fix $\varepsilon \in (0,1)$. Set $\delta = 1 - \frac{\varepsilon}{2}$, then $\delta \in (0,1)$. Using the result stated above, there exists $n_0 \geq 1$ such that

$$0 \leq \int_0^\delta \frac{x^n}{1+x}dx < \frac{\varepsilon}{2} \ ,$$

for any $n \geq n_0$. On the other hand, we have

$$0 \leq \int_\delta^1 \frac{x^n}{1+x}dx \leq \int_\delta^1 dx = 1 - \delta = \frac{\varepsilon}{2} \ .$$

This obviously will imply

$$0 \leq \int_0^1 \frac{x^n}{1+x}dx < \varepsilon \ ,$$

whenever $n \geq n_0$, which proves the intuitive statement

$$\lim_{n \to \infty} \int_0^1 \frac{x^n}{1+x}dx = 0 \ .$$

For (ii), note that

$$\int_0^1 nx(1-x^2)^n dx = \left[-\frac{n}{2(n+1)}(1-x^2)^{n+1}\right]_0^1 = \frac{n}{2(n+1)} \ .$$

This will obviously imply
$$\lim_{n\to\infty} \int_0^1 nx(1-x^2)^n dx = \frac{1}{2}.$$

For (iii), we use the integration-by-parts technique to get
$$\int_1^2 \frac{\sin(nx)}{x} dx = \left[-\frac{\cos(nx)}{nx}\right]_1^2 + \int_1^2 \frac{\cos(nx)}{nx} dx.$$

Hence
$$\int_1^2 \frac{\sin(nx)}{x} dx = \frac{\cos(n)}{n} - \frac{\cos(2n)}{2n} + \frac{1}{n}\int_1^2 \frac{\cos(nx)}{x} dx.$$

Since $\left|\frac{\cos(n)}{n}\right| \leq \frac{1}{n}$, $\left|\frac{\cos(2n)}{2n}\right| \leq \frac{1}{2n}$, and
$$\left|\int_1^2 \frac{\cos(nx)}{x} dx\right| \leq \int_1^2 \left|\frac{\cos(nx)}{x}\right| dx \leq \int_1^2 \frac{1}{x} dx = \ln(2),$$

we have
$$\lim_{n\to\infty} \int_1^2 \frac{\sin(nx)}{x} dx = 0.$$

Solution 7.14

Let us use the substitution $u = \pi - x$. Then we have
$$\int_0^\pi xe^{\sin(x)} dx = \int_\pi^0 -(\pi-u)e^{\sin(\pi-u)} du = \int_0^\pi (\pi-u)e^{\sin(u)} du,$$

which implies
$$\int_0^\pi xe^{\sin(x)} dx = \pi \int_0^\pi e^{\sin(u)} du - \int_0^\pi ue^{\sin(u)} du.$$

Since
$$\int_0^\pi xe^{\sin(x)} dx = \int_0^\pi ue^{\sin(u)} du,$$

we get
$$\int_0^\pi xe^{\sin(x)} dx = \frac{\pi}{2} \int_0^\pi e^{\sin(x)} dx.$$

Solution 7.15

Since $f(x)$ is continuous on $[a,b]$, it is bounded. Set
$$m = \inf\{f(x); \ x \in [a,b]\} \text{ and } M = \sup\{f(x); \ x \in [a,b]\}.$$

Then it is easy to obtain $m(b-a) \leq \int_a^b f(x) dx \leq M(b-a)$, which implies
$$m \leq \frac{1}{b-a} \int_a^b f(x) dx \leq M.$$

CHAPTER 7. INTEGRATION

The Intermediate Value Theorem for continuous functions on closed intervals implies the existence of $c \in [a, b]$ such that
$$\frac{1}{b-a}\int_a^b f(x)dx = f(c).$$
If we fail continuity, then the conclusion does not hold. Indeed, consider
$$f(x) = \begin{cases} 1 & 0 \le x \le 1, \\ 0 & 1 < x \le 2. \end{cases}$$
Then
$$\frac{1}{b-a}\int_a^b f(x)dx = \frac{1}{2}.$$
But $f(x) \ne 1/2$ for all $x \in [0, 2]$.

Solution 7.16

Assume g is differentiable and $u = g(x)$ and $G(u) = \int_0^u f(y)dy$. Set
$$F(x) = G(g(x)) = \int_0^u f(y)dy.$$
Since f is continuous by the Fundamental Theorem of Calculus we have
$$\frac{dG}{du} = \frac{d}{du}\int_0^u f(y)dy = f(u).$$
Using the Chain Rule we obtain
$$F'(x) = G'(g(x)) \cdot g'(x) = f(g(x)) \cdot g'(x).$$
In our problem $g(x) = x^3$, so $g'(x) = 3x^2$, and $F'(x) = f(x^3) \cdot 3x^2$ as claimed.

Solution 7.17

Easy calculations give
$$F(x) = \begin{cases} x & 0 \le x \le 1, \\ 1 & 1 < x \le 2. \end{cases}$$
It is clear that $F(x)$ is continuous and differentiable except at 1, i.e., $F'(1)$ does not exist. Note that we have $F'(x) = f(x)$, for $x \in (0, 2)$ with $x \ne 1$.

Solution 7.18

For any $a, b \in \mathbb{R}$, and $n \ge 1$, we have
$$\int_a^b \sin(nx)dx = \left[-\frac{\cos(nx)}{n}\right]_a^b = \frac{\cos(na)}{n} - \frac{\cos(nb)}{n}.$$
So
$$\lim_{n\to\infty}\int_a^b \sin(nx)dx = \lim_{n\to\infty}\left(\frac{\cos(na)}{n} - \frac{\cos(nb)}{n}\right) = 0.$$

Since this conclusion is true for any $a, b \in \mathbb{R}$, the linearity of the integral will imply

$$\lim_{n \to \infty} \int_a^b S(x) \sin(nx) dx = 0$$

for any step function $S(x)$. In order to get a similar conclusion for Riemann functions, we use the fact that for any Riemann integrable function $f(x)$ and for any $\varepsilon > 0$, there exists a step function $L(x)$ such that $L(x) \leq f(x)$ and $\int_a^b \big(f(x) - L(x)\big) dx \leq \varepsilon$. Since

$$\left| \int_a^b \big(f(x) \sin(nx) - L(x) \sin(nx)\big) dx \right| \leq \int_a^b \big| f(x) \sin(nx) - L(x) \sin(nx) \big| dx ,$$

we get

$$\left| \int_a^b f(x) \sin(nx) - L(x) \sin(nx) dx \right| \leq \int_a^b f(x) - L(x) dx < \varepsilon .$$

It is clear then that the above conclusion also holds for Riemann integrable functions, i.e., for any Riemann integrable function $f(x)$, we have

$$\lim_{n \to \infty} \int_a^b f(x) \sin(nx) dx = 0 .$$

A similar proof will also imply

$$\lim_{n \to \infty} \int_a^b f(x) \cos(nx) dx = 0 ,$$

for any Riemann integrable function $f(x)$. In order to finish the last question, note the following trigonometric identity $2 \sin^2(nx) = 1 - \cos(2nx)$, which implies

$$\int_a^b f(x) \sin^2(nx) dx = \int_a^b f(x) \frac{1 - \cos(2nx)}{2} dx = \frac{1}{2} \int_a^b f(x) dx - \frac{1}{2} \int_a^b f(x) \cos(2nx) dx .$$

The previous conclusions will then imply

$$\lim_{n \to \infty} \int_a^b f(x) \sin^2(nx) dx = \frac{1}{2} \int_a^b f(x) dx .$$

Solution 7.19

Assume not, i.e., $f(x_0) \neq 0$. Then we have $f^2(x_0) > 0$. Since $f^2(x)$ is also continuous at x_0, then there exists $\delta > 0$ such that for any $x \in (x_0 - \delta, x_0 + \delta) \subset (a, b)$, we have $f^2(x) \geq \dfrac{f^2(x_0)}{2}$. Since

$$\int_{x_0 - \delta}^{x_0 + \delta} f^2(x) dx \leq \int_a^b f^2(x) dx$$

and

$$\int_{x_0 - \delta}^{x_0 + \delta} \frac{f^2(x_0)}{2} dx \leq \int_{x_0 - \delta}^{x_0 + \delta} f^2(x) dx ,$$

CHAPTER 7. INTEGRATION

we get
$$\delta f^2(x_0) = \int_{x_0-\delta}^{x_0+\delta} \frac{f^2(x_0)}{2} dx \leq \int_a^b f^2(x) dx .$$

This obviously contradicts our assumption $\int_a^b f^2(x) dx = 0$. From this conclusion, we conclude that the set $\{x \in [a,b]; f(x) \neq 0\}$ is a subset of all discontinuous points of $f(x)$. Since $f(x)$ is Riemann integrable, this set must be negligible or have measure 0.

Solution 7.20

a) Note the inequality
$$\frac{1}{\sqrt{1+x^3}} \leq \frac{1}{\sqrt{x^3}} ,$$
for any $x \geq 1$. Since $\int_1^\infty \frac{1}{\sqrt{x^3}} dx = \int_1^\infty \frac{1}{x^{3/2}} dx$ is convergent (because $3/2 > 1$), the basic comparison test will force $\int_1^\infty \frac{1}{\sqrt{1+x^3}} dx$ to be convergent.

b) We use the integration by parts to get
$$\int_1^A \frac{\sin(x)}{x} dx = \left[-\frac{\cos(x)}{x}\right]_1^A - \int_1^A \frac{\cos(x)}{x^2} dx ,$$
for any $A > 1$. Since
$$\lim_{A \to \infty} \left[-\frac{\cos(x)}{x}\right]_1^A = \frac{\cos(1)}{1} = \cos(1) ,$$
and the improper integral $\int_1^\infty \frac{\cos(x)}{x^2} dx$ is absolutely convergent (because $\left|\frac{\cos(x)}{x^2}\right| \leq \frac{1}{x^2}$ and $\int_1^\infty \frac{1}{x^2} dx$ is convergent), we conclude that $\int_1^\infty \frac{\sin(x)}{x} dx$ is convergent. Let us show that it is not absolutely convergent. Indeed assume not, then $\int_1^\infty \left|\frac{\sin(x)}{x}\right| dx$ is convergent. Since
$$\int_1^\infty \left|\frac{\sin(x)}{x}\right| dx = \sum_{n=1}^\infty \int_{n\pi}^{(n+1)\pi} \left|\frac{\sin(x)}{x}\right| dx,$$
we conclude that the series on the left side is convergent. But
$$\int_{n\pi}^{(n+1)\pi} \left|\frac{\sin(x)}{x}\right| dx = \int_0^\pi \frac{|\sin(x+n\pi)|}{x+n\pi} dx = \int_0^\pi \frac{|(-1)^n \sin(x)|}{x+n\pi} dx = \int_0^\pi \frac{\sin(x)}{x+n\pi} dx .$$
Since
$$\int_0^\pi \frac{\sin(x)}{x+n\pi} dx \geq \int_0^\pi \frac{\sin(x)}{\pi+n\pi} dx = \frac{2}{(n+1)\pi} ,$$
the basic comparison test for positive series will then force the series $\sum_{n=1}^\infty \frac{2}{(n+1)\pi}$ is convergent. Contradiction.

c) Note that $\int_0^\infty \dfrac{x}{1+x^2\sin^2(x)}dx$ is convergent if and only if $\int_1^\infty \dfrac{x}{1+x^2\sin^2(x)}dx$ is convergent. Since

$$\frac{1}{2x} \leq \frac{x}{1+x^2} \leq \frac{x}{1+x^2\sin^2(x)}$$

for any $x \geq 1$ and $\int_1^\infty \dfrac{1}{2x}dx$ is divergent, the basic comparison test will then force $\int_1^\infty \dfrac{x}{1+x^2\sin^2(x)}dx$ to be divergent. Hence the improper integral $\int_0^\infty \dfrac{x}{1+x^2\sin^2(x)}dx$ is divergent.

Solution 7.21

Set $\mu = \dfrac{1+\alpha}{2}$. If $\alpha > 1$, then $1 < \mu < \alpha$. Since $\lim_{x\to\infty} \dfrac{x^\mu}{x^\alpha \ln^\beta(x)} = 0$, for any $\beta \in \mathbb{R}$, there exists $A > 0$ such that for any $x \geq A$, we have $\dfrac{x^\mu}{x^\alpha \ln^\beta(x)} \leq 1$ which implies $\dfrac{1}{x^\alpha \ln^\beta(x)} \leq \dfrac{1}{x^\mu}$. Since the improper integral $\int_1^\infty \dfrac{1}{x^\mu}dx$ is convergent, the basic comparison test will force $\int_2^\infty \dfrac{1}{x^\alpha \ln^\beta(x)}dx$ to be convergent. If $\alpha < 1$, then $1 > \mu > \alpha$. Since $\lim_{x\to\infty} \dfrac{x^\mu}{x^\alpha \ln^\beta(x)} = \infty$, for any $\beta \in \mathbb{R}$, there exists $A > 0$ such that for any $x \geq A$, we have $\dfrac{x^\mu}{x^\alpha \ln^\beta(x)} \geq 1$ which implies $\dfrac{1}{x^\alpha \ln^\beta(x)} \geq \dfrac{1}{x^\mu}$. Since the improper integral $\int_1^\infty \dfrac{1}{x^\mu}dx$ is divergent, the basic comparison test will force $\int_2^\infty \dfrac{1}{x^\alpha \ln^\beta(x)}dx$ to be divergent. Finally, assume $\alpha = 1$. Then for any $A > 2$, we have

$$\int_2^A \frac{1}{x \ln^\beta(x)}dx = \int_{\ln(2)}^{\ln(A)} \frac{1}{x^\beta}dx .$$

Since $\int_{\ln(2)}^\infty \dfrac{1}{x^\beta}dx$ is convergent if and only if $\beta > 1$, $\int_2^\infty \dfrac{1}{x \ln^\beta(x)}dx$ is convergent if and only if $\beta > 1$.

Solution 7.22

Set $F(t) = \int_a^t f(x)dx$. It is clear that the improper integral $\int_a^\infty f(x)dx$ converges if and only if $\lim_{t\to\infty} F(t)$ exists. Assume that the improper integral converges. Then for any $\varepsilon > 0$, there exists $A > 0$ such that for any $t > A$ we have

$$\left| F(t) - \int_a^\infty f(x)dx \right| < \frac{\varepsilon}{2} .$$

Hence for any $t_1, t_2 > A$, we have

$$\left| F(t_1) - F(t_2) \right| \leq \left| F(t_1) - \int_a^\infty f(x)dx \right| + \left| F(t_2) - \int_a^\infty f(x)dx \right| < \frac{\varepsilon}{2} + \frac{\varepsilon}{2} = \varepsilon .$$

Assume the converse is true, i.e., for every $\varepsilon > 0$ there exists $A > a$ such that for any $t_1, t_2 > A$ we have

$$\left| \int_{t_1}^{t_2} f(x)dx \right| < \varepsilon .$$

Let us prove that the improper integral $\int_a^\infty f(x)dx$ is convergent. Note that $\lim_{t\to\infty} F(t)$ exists if and only if for any sequence $\{t_n\}$ which goes to ∞, the sequence $\{F(t_n)\}$ is convergent. In $\mathbb{R}$ convergence of sequences is equivalent to the Cauchy behavior. Hence $\lim_{t\to\infty} F(t)$ exists if and only if for any sequence $\{t_n\}$ which goes to ∞, the sequence $\{F(t_n)\}$ is Cauchy. Our assumption forces this to be true. Indeed, let $\{t_n\}$ be a sequence which goes to 0. Let $\varepsilon > 0$. Then there exists $A > 0$ such that for any $t_1, t_2 > A$ we have

$$\left| F(t_1) - F(t_2) \right| = \left| \int_{t_1}^{t_2} f(x)dx \right| < \varepsilon .$$

Since $\{t_n\}$ goes to ∞, there exists $n_0 \geq 1$ such that for any $n \geq n_0$ we have $t_n > A$. So for any $n, m \geq n_0$, we have

$$\left| F(t_n) - F(t_m) \right| = \left| \int_{t_n}^{t_m} f(x)dx \right| < \varepsilon ,$$

which translates into $\{F(t_n)\}$ being a Cauchy sequence.

Solution 7.23

Let us use the Cauchy criteria proved in the previous problem to prove our claim. Let $\varepsilon > 0$. Since $g(x)$ is bounded, there exists $M > 0$ such that $|g(x)| \leq M$ for all $x \in [a, \infty)$. Since $\int_a^\infty f(x)dx$ is convergent, there exists $A > 0$ such that for any $t_1, t_2 > A$ we have

$$\left| \int_{t_1}^{t_2} f(x)dx \right| < \frac{\varepsilon}{2M} .$$

By the Second Mean Value Theorem for integrals, and for any $t_1, t_2 > A$, there exists c between t_1 and t_2 such that

$$\int_{t_1}^{t_2} f(x)g(x)dx = g(t_1) \int_{t_1}^{c} f(x)dx + g(t_2) \int_{c}^{t_2} f(x)dx .$$

Hence

$$\left| \int_{t_1}^{t_2} f(x)g(x)dx \right| \leq |g(t_1)| \left| \int_{t_1}^{c} f(x)dx \right| + |g(t_2)| \left| \int_{c}^{t_2} f(x)dx \right| .$$

But

$$\left| \int_{t_1}^{c} f(x)dx \right| < \frac{\varepsilon}{2M} \text{ and } \left| \int_{c}^{t_2} f(x)dx \right| < \frac{\varepsilon}{2M} ,$$

which forces the inequality

$$\left| \int_{t_1}^{t_2} f(x)g(x)dx \right| < \frac{\varepsilon}{2M}|g(t_1)| + \frac{\varepsilon}{2M}|g(t_2)| < \varepsilon$$

to be true.

Solution 7.24

The lower integral of any function is well defined provided the function is bounded. Therefore for any bounded function $f(x)$, and by definition of the lower integral, for any $\varepsilon > 0$, there exists a step function $L(x) \leq f(x)$ such that

$$\underline{\int_a^b} \Big(f(x) - L(x)\Big) dx < \frac{\varepsilon}{2}.$$

Also note the existence of a continuous function $c(x) \leq L(x)$ such that

$$\underline{\int_a^b} \Big(L(x) - c(x)\Big) dx < \frac{\varepsilon}{2}.$$

Putting all this together, we conclude that for any bounded function $f(x)$, and for any $\varepsilon > 0$, there exists a continuous function $c(x) \leq f(x)$ such that

$$\underline{\int_a^b} \Big(f(x) - c(x)\Big) dx < \varepsilon.$$

Note that if $f(x) \geq 0$, then the construction of $c(x)$ will be done to have $c(x) \geq 0$ as well. Back to our claim. Let $\varepsilon > 0$. Then there exists a positive continuous function $c_1(x) \leq f_1(x)$ such that

$$\underline{\int_a^b} \Big(f_1(x) - c_1(x)\Big) dx < \frac{\varepsilon}{2^2}.$$

Since $\min(c_1(x), f_2(x))$ is bounded and positive, there exists a positive continuous function $c_2(x) \leq \min(c_1(x), f_2(x))$ such that

$$\underline{\int_a^b} \Big(\min(c_1(x), f_2(x)) - c_2(x)\Big) dx < \frac{\varepsilon}{2^3}.$$

Since $f_2(x) - c_2(x) \leq f_2(x) - \min(c_1(x), f_2(x)) + \min(c_1(x), f_2(x)) - c_2(x)$, and $f_2(x) \leq f_1(x)$, we get

$$\underline{\int_a^b} \Big(f_2(x) - c_2(x)\Big) dx \leq \underline{\int_a^b} \Big(f_1(x) - c_1(x)\Big) dx + \underline{\int_a^b} \Big(\min(c_1(x), f_2(x)) - c_2(x)\Big) dx < \frac{\varepsilon}{2^2} + \frac{\varepsilon}{2^3}.$$

By the induction argument, a similar construction will lead to the existence of a decreasing sequence of positive continuous functions $\{c_n(x)\}$ such that $c_n(x) \leq f_n(x)$ and

$$\underline{\int_a^b} \Big(f_n(x) - c_n(x)\Big) dx < \frac{\varepsilon}{2^2} + \frac{\varepsilon}{2^3} + \cdots + \frac{\varepsilon}{2^{n+1}} < \frac{\varepsilon}{2}.$$

Since $\{f_n(x)\}$ converges pointwise to 0 on $[a, b]$ this will force the sequence $\{c_n(x)\}$ to also converge pointwise to 0 on $[a, b]$. Dini's theorem will imply that $\{c_n(x)\}$ converges uniformly to 0 on $[a, b]$. So there exists $n_0 \geq 1$ such that for any $n \geq n_0$ we have $c_n(x) \leq \dfrac{\varepsilon}{2(b-a)}$ for any $x \in [a, b]$. Hence

$$\underline{\int_a^b} f_n(x) dx \leq \underline{\int_a^b} \Big(f_n(x) - c_n(x)\Big) dx + \int_a^b c_n(x) dx < \frac{\varepsilon}{2} + \frac{\varepsilon}{2(b-a)}(b-a) = \varepsilon.$$

CHAPTER 7. INTEGRATION

whenever $n \geq n_0$. This finishes the proof of our claim.

Solution 7.25

Since $f(x)$ is Riemann integrable, $\{f_n(x) - f(x)\}$ is a sequence of Riemann integrable functions which decreases to 0. Obviously they are all bounded functions. The Luxemburg Monotone Convergence Theorem will imply

$$\lim_{n \to \infty} \underline{\int_a^b} \Big(f_n(x) - f(x)\Big) dx = 0 \;.$$

But $\underline{\int_a^b} \Big(f_n(x) - f(x)\Big) dx = \int_a^b \Big(f_n(x) - f(x)\Big) dx$, hence

$$\lim_{n \to \infty} \int_a^b \Big(f_n(x) - f(x)\Big) dx = 0 \;,$$

which implies

$$\lim_{n \to \infty} \int_a^b f_n(x) = \int_a^b f(x) dx \;.$$

Solution 7.26

Set $\widehat{f_n}(x) = |f_n(x) - f(x)|$. Our assumptions imply that $\{\widehat{f_n}(x)\}$ converges to 0 pointwise. Set $h_n(x) = \sup_{k \geq n} \widehat{f_k}(x)$. Clearly $\{h_n(x)\}$ also converges pointwise to 0 on $[a,b]$. But these functions are not necessarily Riemann integrable. But they are bounded. The Luxemburg Monotone Convergence Theorem will force the conclusion

$$\lim_{n \to \infty} \underline{\int_a^b} h_n(x) dx = 0 \;.$$

Since $0 \leq \widehat{f_n}(x) \leq h_n(x)$, we get

$$\lim_{n \to \infty} \underline{\int_a^b} \widehat{f_n}(x) dx = 0 \;.$$

But $\widehat{f_n}$ are Riemann integrable which implies

$$\lim_{n \to \infty} \int_a^b \widehat{f_n}(x) dx = 0 \;.$$

Since $\left|\int_a^b \Big(f_n(x) - f(x)\Big) dx\right| \leq \int_a^b \widehat{f_n}(x) dx$, we get

$$\lim_{n \to \infty} \int_a^b \Big(f_n(x) - f(x)\Big) dx = 0 \;,$$

or

$$\lim_{n \to \infty} \int_a^b f_n(x) dx = \int_a^b f(x) dx.$$

Solution 7.27

If $\liminf_{n\to\infty} \int_a^b f_n(x)dx = \infty$, then the conclusion is obvious. Assume that $\liminf_{n\to\infty} \int_a^b f_n(x)dx < \infty$. Then there exists a subsequence $\{f_{n_k}\}$ of $\{f_n\}$ such that

$$\lim_{n_k\to\infty} \int_a^b f_{n_k}(x)dx = \liminf_{n\to\infty} \int_a^b f_n(x)dx \ .$$

Clearly the subsequence $\{f_{n_k}\}$ also converges pointwise to $f(x)$. Set $h_{n_i}(x) = \inf_{n_k \geq n_i} f_{n_k}(x)$. Then $\{h_{n_k}(x)\}$ also converges pointwise to $f(x)$ and is increasing. It is easy to see that this sequence is bounded. The Luxemburg Monotone Convergence Theorem applied to $\{f(x) - h_{n_k}(x)\}$ will easily imply that

$$\lim_{n\to\infty} \int_a^b h_{n_k}(x)dx = \int_a^b f(x)dx \ .$$

Since $h_{n_k}(x) \leq f_{n_k}(x)$, we get

$$\int_a^b f(x)dx \leq \lim_{n\to\infty} \int_a^b f_{n_k}(x)dx = \liminf_{n\to\infty} \int_a^b f_n(x)dx.$$

Solution 7.28

a) If T were a contraction, then there exist $0 < \lambda < 1$ such that

$$\| Tf - Tg \| \leq \lambda \| f - g \| \quad \forall f, g \in C[0,1].$$

Taking $f(t) = 1$, $g(t) = 0$ in $C[0,1]$, we have

$$\| Tf - Tg \| = \sup | \int_0^x dt - \int_0^x 0\, dt | = \sup_{0 \leq x \leq 1} |x| = 1,$$

$$\| f - g \| = \sup_{0 \leq x \leq 1} |1 - 0| = 1$$

and hence we will have $1 \leq \lambda \cdot 1$ which is a contraction since $\lambda < 1$. Therefore T is not a contraction.

b) Consider $f(x) = 0$, then $Tf = f$, therefore we have the existence of the fixed point. To show the uniqueness of the fixed point, assume not, suppose we have another fixed point say h such that $Th = h$ or equivalently

$$\int_0^x h(t)\, dt = h(x).$$

From the Fundamental Theorem of Calculus we have $\dfrac{dh}{dx} = h(x)$, and the solution to this differential equation is $h(x) = Ce^x$. Since $h(0) = 0$, we have $C = 0$ and therefore $h(x) = 0 = f(x)$.

CHAPTER 7. INTEGRATION

c) First observe that
$$T^2 f(x) = T(Tf(x)) = \int_0^x \left(\int_0^t f(s)\, ds \right) dt$$

and

$$\| T^2 f - T^2 g \| = \sup_{0 \leq x \leq 1} \left| \int_0^x \left(\int_0^t \big(f(s) - g(s)\big) ds \right) dt \right| \leq \sup_{0 \leq x \leq 1} \int_0^x \left(\int_0^t |f(s) - g(s)|\, ds \right) dt,$$

$$\| T^2 f - T^2 g \| \leq \| f - g \| \sup_{0 \leq x \leq 1} \int_0^x t\, dt = \| f - g \| \sup_{0 \leq x \leq 1} \frac{x^2}{2} = \frac{1}{2} \| f - g \|.$$

We showed:
$$\| T^2 f - T^2 g \| \leq \frac{1}{2} \| f - g \|$$

and hence T^2 is a contraction.

Solution 7.29

First observe that the given initial value problem
$$\frac{dy}{dx} = 3xy, \quad y(0) = 1$$

can be written as
$$f(x) = 1 + \int_0^x 3s\, f(s)\, ds,$$

since $f(0) = 1$ and $\dfrac{dy}{dx} = 3x f(x)$ by the fundamental theorem of calculus. Using the Picard iteration form the following sequence:

$f_0(x) = 1$

$f_1(x) = T(f_0(x)) = 1 + \int_0^x 3s \cdot 1\, ds = 1 + (3/2)s^2 |_0^x = 1 + (3/2)x^2$

$f_2(x) = T(f_1(x)) = 1 + \int_0^x 3s \cdot \left(1 + \frac{3}{2}s^2\right) ds = 1 + 3\frac{s^2}{2} + (3/2)\frac{s^4}{4}|_0^x = 1 + \left(\frac{3}{2}x^2\right) + \left(\frac{3}{2}x^2\right)^2 \frac{1}{2!}.$

It is clear that we have

$$f_n(x) = T(f_{n-1}(x)) = 1 + \left(\frac{3}{2}x^2\right) + \left(\frac{3}{2}x^2\right)^2 \frac{1}{2!} + \cdots + \left(\frac{3}{2}x^2\right)^n \frac{1}{n!} = \sum_{k=0}^n \frac{(\frac{3}{2}x^2)^n}{n!}.$$

But this sequence $f_n(x)$ converges to $f(x) = e^{\frac{3}{2}x^2}$.

Solution 7.30

Given the initial value problem is $f'(x) = 1 + x - f(x)$ for $-1/2 \leq x \leq 1/2$ under $f(0) = 1$. So we have the integral equation

$$f(x) = 1 + \int_0^x (1 + t - f(t))\, dt = 1 + x + \frac{x^2}{2} - \int_0^x f(t)\, dt.$$

So define $T : C[-\frac{1}{2}, \frac{1}{2}] \to C[-\frac{1}{2}, \frac{1}{2}]$ by

$$Tf(x) = 1 + x + \frac{x^2}{2} - \int_0^x f(t)dt$$

$$|Tf(x) - Tg(x)| = \left|\int_0^x f(t) - g(t)dt\right| \leq \int_0^x |f(t) - g(t)|dt$$

$$\leq \|f - g\|_\infty \int_0^x dt \leq \frac{1}{2}\|f - g\|_\infty.$$

This estimate is independent of $x \in [-1/2, 1/2]$, so we have $\|Tf - Tg\|_\infty \leq 1/2\|f - g\|_\infty$.

$$f_1(x) = Tf_0(x) = 1 + x + \frac{x^2}{2} - \int_0^x 1 dt = 1 + \frac{1}{2}x^2.$$

$$f_2(x) = Tf_1(x) = 1 + x + \frac{x^2}{2} - \int_0^x \left(1 + \frac{1}{2}t^2\right)dt = 1 + \frac{1}{2}x^2 - \frac{1}{6}x^3.$$

$$f_3(x) = Tf_2(x) = 1 + x + \frac{x^2}{2} - \int_0^x \left(1 + \frac{1}{2}x^2 - \frac{1}{6}x^3\right)dt = 1 + \frac{1}{2}x^2 - \frac{1}{6}x^3 + \frac{1}{24}x^4.$$

It is easy to show by induction that

$$f_n(x) = 1 + \frac{1}{2!}x^2 - \frac{1}{3!}x3 + \frac{1}{4!}x^4 - \frac{1}{5!}x^5 + \cdots + \frac{1}{(n+1)!}(-x)^{n+1}.$$

Clearly this sequence converges to $e^{-x} + x$.

Solution 7.31

Let $F(x) = \int_0^\pi \frac{\sin xt}{t} dt$ and notice that $f(x,t) = \sin xt$ is a continuous function on $[a,b] \times [0,\pi]$. Let x_0 be a fixed x-value, then given $\varepsilon > 0$ choose $\delta = \frac{\varepsilon}{\pi}$, then

$$|F(x) - F(x_0)| = \left|\int_0^\pi \frac{\sin xt}{t} dt - \int_0^\pi \frac{\sin x_0 t}{t} dt\right| \leq \int_0^\pi \frac{|\sin xt - \sin x_0 t|}{t} dt$$

$$\leq \int_0^\pi \frac{|xt - x_0 t|}{t} dt = \pi|x - x_0| \leq \pi\delta = \varepsilon,$$

thus $|F(x) - F(x_0)| < \varepsilon$, and F is continuous.

Solution 7.32

1. In order to prove the existence of such polynomials, we will need Euler's formula, i.e., $e^{i\theta} = \cos(\theta) + i\sin(\theta)$. Indeed we have

$$e^{in\theta} = \cos(n\theta) + i\sin(n\theta) = \left(e^{i\theta}\right)^n.$$

CHAPTER 7. INTEGRATION

But
$$\left(e^{i\theta}\right)^n = \sum_{k=0}^{n} \binom{n}{k} i^k \sin^k(\theta) \cos^{n-k}(\theta).$$

So the real part of the two complex numbers must be equal which gives
$$\cos(n\theta) = \sum_{0 \leq 2k \leq n} \binom{n}{2k}(-1)^k \sin^{2k}(\theta) \cos^{n-2k}(\theta).$$

But $\sin^{2k}(\theta) = \left(\sin^2(\theta)\right)^k = \left(1 - \cos^2(\theta)\right)^k$, which implies
$$\cos(n\theta) = \sum_{0 \leq 2k \leq n} \binom{n}{2k}(-1)^k \left(1 - \cos^2(\theta)\right)^k \cos^{n-2k}(\theta).$$

Set
$$T_n(x) = \sum_{0 \leq 2k \leq n} \binom{n}{2k}(-1)^k \left(1 - x^2\right)^k x^{n-2k}.$$

Then $T_n(\cos(\theta)) = \cos(n\theta)$. Clearly T_n is a polynomial function with degree n.

2. We have the trigonometric identity
$$\cos(n+1)x + \cos(n-1)x = 2\cos(x)\cos(nx).$$

Hence
$$T_{n+1}(\cos(x)) + T_{n-1}(\cos(x)) = 2\cos(x)T_n(\cos(x))$$

which implies $T_{n+1} = 2xT_n - T_{n-1}$, for any $n \geq 1$.

3. In order to find the integrals
$$\int_{-1}^{1} \frac{T_n(x)T_m(x)}{\sqrt{1-x^2}} dx, \ n, m \in \mathbb{N},$$

we use the change of variable $x = \cos(t)$. Hence
$$\int_{-1}^{1} \frac{T_n(x)T_m(x)}{\sqrt{1-x^2}} dx = \int_{\pi}^{0} \frac{T_n(\cos(t))T_m(\cos(t))}{\sqrt{1-\cos^2(t)}}(-\sin(t))dt = \int_0^{\pi} T_n(\cos(t))T_m(\cos(t))dt.$$

Using the main property of the polynomial functions T_n, we get
$$\int_{-1}^{1} \frac{T_n(x)T_m(x)}{\sqrt{1-x^2}} dx = \int_0^{\pi} \cos(nt)\cos(mt)dt.$$

If $n \neq m$, we use the identity
$$\cos(nt)\cos(mt) = \frac{1}{2}\Big(\cos(n-m)t + \cos(n+m)t\Big),$$

to get
$$\int_0^{\pi} \cos(nt)\cos(mt)dt = \frac{1}{2}\left[\frac{\sin(n-m)t}{n-m} + \frac{\sin(n+m)t}{n+m}\right]_0^{\pi} = 0.$$

If $n = m$, then
$$\int_{-1}^{1} \frac{T_n^2(x)}{\sqrt{1-x^2}} dx = \int_0^\pi \cos^2(nt) dt = \frac{1}{2} \int_0^\pi \Big(1 + \cos(2nt)\Big) dt = \frac{\pi}{2},$$
if $n \neq 0$. If $n = 0$, then
$$\int_{-1}^{1} \frac{T_0^2(x)}{\sqrt{1-x^2}} dx = \pi.$$

Chapter 8

Series

That fondness for science, ... that affability and condescension which God shows to the learned, that promptitude with which he protects and supports them in the elucidation of obscurities and in the removal of difficulties, has encouraged me to compose a short work on calculating by al-jabr and al-muqabala, confining it to what is easiest and most useful in arithmetic. [al-jabr means "restoring," referring to the process of moving a subtracted quantity to the other side of an equation; al-muqabala is "comparing" and refers to subtracting equal quantities from both sides of an equation.]

Musa Al-Khwarizmi (about 790–about 840)

- Let (a_k) be a sequence of real numbers. We use the notation $s_n = \sum_{k=0}^{n} a_k$ to denote the nth *partial sum* of the infinite series $s_\infty = \sum_{k=0}^{\infty} a_k$. If the sequence of partial sums (s_n) converges to a real number s, we say that the series $\sum_{k} a_k$ is *convergent* and we write $s = \sum_{k=0}^{\infty} a_k$. A series that is not convergent is called *divergent*.

- An infinite series $\sum_{k=0}^{\infty} a_k$ is said to *converge absolutely* if $\sum_{k=0}^{\infty} |a_k|$ converges. If a series converges absolutely, then it converges. Furthermore, an absolutely convergent series converges to the same sum in whatever order the terms are taken.

- If $\sum_{k=0}^{\infty} a_k$ converges but $\sum_{k=0}^{\infty} |a_k|$ diverges, then we say $\sum_{k=0}^{\infty} a_k$ *converges conditionally*. Any conditionally convergent series can be rearranged to obtain a series which converges to any given sum or diverges to ∞ or $-\infty$.

- *Ratio Test*: Given a series $\sum_{k=1}^{\infty} a_k$ with $a_k \neq 0$, if a_k satisfies
$$\lim_{k \to \infty} |\frac{a_{k+1}}{a_k}| = r < 1,$$
then the series converges absolutely.

- *Root Test*: Let $\sum_{k=1}^{\infty} a_k$ be a series and
$$\alpha := \overline{\lim_{k \to \infty}} \sqrt[k]{|a_k|}.$$
Then the following hold:

  a) $\sum_{k=1}^{\infty} a_k$ converges absolutely if $\alpha < 1$.

  b) $\sum_{k=1}^{\infty} a_k$ diverges if $\alpha > 1$.

  For $\alpha = 1$ both convergence and divergence of $\sum_{k=1}^{\infty} a_k$ are possible.

- *Cauchy Criterion for Series*: The series $\sum_{k=1}^{\infty} a_k$ converges if and only if given $\varepsilon > 0$, there exists $N \in \mathbb{N}$ such that whenever $n > m \geq N$ it follows that
$$|a_{m+1} + a_{m+2} + \cdots + a_n| < \varepsilon.$$

- *Comparison Test*: Assume (a_k) and (b_k) are sequences satisfying $0 \leq a_k \leq b_k$ for all $k \in \mathbb{N}$.

  a) If $\sum_{k=1}^{\infty} b_k$ converges, then $\sum_{k=1}^{\infty} a_k$ converges.

  b) If $\sum_{k=1}^{\infty} a_k$ diverges, then $\sum_{k=1}^{\infty} b_k$ diverges.

- *Geometric Series*: A series is called *geometric* if it is of the form
$$\sum_{k=1}^{\infty} ar^k = a + ar + ar^2 + ar^3 \cdots$$
and
$$\sum_{k=1}^{\infty} ar^k = \frac{a}{1-r}$$
if and only if $|r| < 1$. In case $r = 1$ and $a \neq 0$, the series diverges.

CHAPTER 8. SERIES

- *Alternating Series Test*: Let (a_n) be a sequence satisfying

  a) $a_1 \geq a_2 \geq a_3 \geq \cdots \geq a_n \geq a_{n+1} \geq \cdots$ and

  b) $(a_n) \to 0$.

  Then the alternating series $\sum_{n=1}^{\infty}(-1)^{n+1}a_n$ converges.

- Let $\sum_{k=1}^{\infty} a_k$ be a series. A *rearrangement* is the series $\sum_{k=1}^{\infty} a_{\sigma(k)}$ where σ is a permutation of $\{1,2,3,\ldots\}$. The summands of the rearrangement $\sum_{k=1}^{\infty} a_{\sigma(k)}$ are the same as those of the original series, but they occur in different order. If σ is a permutation of $\mathbb{N}$ with $\sigma(k) = k$ for almost all $k \in \mathbb{N}$, then $\sum_{k=1}^{\infty} a_k$ and $\sum_{k=1}^{\infty} a_{\sigma(k)}$ have the same convergence behavior, and their values are equal if the series converge. For a permutation $\sigma(k) \neq k$ for infinitely many $k \in \mathbb{N}$, this may not be true.

- If $\sum_{k=1}^{\infty} a_k$ converges absolutely, then any rearrangement of this series converges to the same limit.

Problem 8.1 Prove that if the series $\sum_{k=1}^{\infty} a_k$ converges, then $\{a_k\}$ is a null sequence.

Problem 8.2 Find the infinite series and its sum if the sequence $\{s_n\}$ of partial sums is given by
$$\{s_n\} = \left\{\frac{n+1}{n}\right\}_{n \in \mathbb{N}}.$$

Problem 8.3 Show that the Euler's series $\sum_{k=1}^{\infty} \frac{1}{k^2}$ converges. Find the sum of the series.

Problem 8.4 Show that the harmonic series $\sum_{k=1}^{\infty} \frac{1}{k}$ diverges.

Problem 8.5 Discuss the convergence of the series $\sum_{n=2}^{\infty} \dfrac{1}{n \ln^p(n)}$ depending on $p \geq 0$. These series are known as the Bertrand series.

Problem 8.6 Find the sum of the series

a) $\displaystyle\sum_{k=1}^{\infty} \dfrac{1}{k^2 + k}$

b) $\displaystyle\sum_{k=1}^{\infty} \dfrac{\ln\left(\dfrac{k^{k+1}}{(k+1)^k}\right)}{k(k+1)}$

c) $\displaystyle\sum_{k=1}^{\infty} \dfrac{1}{(2k-1)^2}$

Problem 8.7 Suppose that $\sum x_n$ is a series of positive terms which is convergent. Show that $\sum \dfrac{1}{x_n}$ is divergent. What about the converse?

Problem 8.8 Suppose that $\sum x_n$ is a series of positive terms which is convergent. Show that $\sum x_n^2$ and $\sum \sqrt{x_n x_{n+1}}$ are convergent.

Problem 8.9 Suppose that $\sum x_n$ is a series of positive terms which is convergent. Show that $\sum \dfrac{\sqrt{x_n}}{n}$ is convergent.

Problem 8.10 Show that the series $\sum \dfrac{(-1)^n}{\sqrt{n} + (-1)^n}$ is divergent while $\sum \dfrac{(-1)^n}{\sqrt{n}}$ is convergent. Deduce from this that the limit convergence test does not work for nonpositive series.

Problem 8.11 Discuss the convergence or divergence of $\sum x_n$ where

$$x_n = \int_1^{\infty} e^{-x^n} dx, \ n = 1, \dots .$$

CHAPTER 8. SERIES

Problem 8.12 Let $\{x_n\}$ and $\{\varepsilon_n\}$ be two sequences of real numbers such that

1. the sequence of partial sums $\{s_n\}$ of $\sum x_n$ is bounded, i.e., there exists $M > 0$ such that
$$|s_n| = |x_1 + \cdots + x_n| \leq M, \ n = 1, \ldots;$$

2. $\lim_{n \to \infty} \varepsilon_n = 0$;

3. the series $\sum |\varepsilon_{n+1} - \varepsilon_n|$ is convergent.

Then the series $\sum \varepsilon_n x_n$ is convergent. This conclusion is known as *Abel's test* or *Abel's theorem*.

Problem 8.13 Show that for any $n \geq 1$, we have
$$1 + \frac{1}{2!} + \cdots + \frac{1}{n!} \leq e \leq 1 + \frac{1}{2!} + \cdots + \frac{1}{n(n!)}.$$

Use these inequalities to discuss the convergence or divergence of
$$\sum_{n \geq 1} \sin(\pi e n!).$$

Problem 8.14 Consider the sequence
$$x_n = \frac{(-1)^n}{p\,n + 1}$$

where $p > 0$. Show that $\sum x_n$ is convergent and its sum is
$$\sum_{n=0}^{\infty} x_n = \int_0^1 \frac{dt}{1 + t^p}.$$

In particular, show
$$\sum_{n=0}^{\infty} \frac{(-1)^n}{n+1} = \ln 2, \quad \sum_{n=0}^{\infty} \frac{(-1)^n}{2n+1} = \frac{\pi}{4}, \quad \text{and} \quad \sum_{n=0}^{\infty} \frac{(-1)^n}{3n+1} = \frac{1}{3}\left(\ln(2) + \frac{\pi}{\sqrt{3}}\right).$$

Problem 8.15 Show that the series
$$\sum_{n \geq 1} \frac{\cos(n\theta)}{n} \quad \text{and} \quad \sum_{n \geq 1} \frac{\sin(n\theta)}{n}$$

are convergent, where $0 < \theta < 2\pi$.

Problem 8.16 (Dirichlet's Rearrangement Theorem) Let $\sum x_n$ be an absolutely convergent series, with $\sum_{n=1}^{\infty} x_n = s$, and let $\sum y_n$ be any rearrangement of $\sum x_n$. Show that $\sum y_n$ converges, and $\sum_{n=1}^{\infty} y_n = s$.

Problem 8.17 Given two series $\sum x_n$ and $\sum y_n$, define $z_n = \sum_{k=0}^{n} x_k y_{n-k}$. Suppose that $\sum x_n$ and $\sum y_n$ are absolutely convergent. Show that $\sum z_n$ is absolutely convergent, and

$$\sum z_n = \sum x_n \cdot \sum y_n.$$

Problem 8.18 Consider the positive series $\sum x_n$ and $\sum y_n$ with $x_n > 0$ and $y_n > 0$. Assume that there exists $N \geq 1$ such that

$$\frac{x_{n+1}}{x_n} \leq \frac{y_{n+1}}{y_n}$$

for $n \geq N$. Show that if $\sum x_n$ is divergent, then $\sum y_n$ is divergent as well.

Problem 8.19 Consider the positive series $\sum x_n$ with $x_n > 0$. Assume that there exists $N \geq 1$ and $p > 1$ is a real number such that

$$n\left(1 - \frac{x_{n+1}}{x_n}\right) \geq p$$

for any $n \geq N$. Show that $\sum x_n$ is convergent.

Problem 8.20 Consider the positive series $\sum x_n$ with $x_n > 0$. Assume that there exists $N \geq 1$ such that

$$n\left(1 - \frac{x_{n+1}}{x_n}\right) \leq 1$$

for any $n \geq N$. Show that $\sum x_n$ is divergent.

Problem 8.21 Consider the positive series $\sum x_n$ with $x_n > 0$. Show that

(a) if $\lim_{n \to \infty} n\left(1 - \frac{x_{n+1}}{x_n}\right) > 1$, then $\sum x_n$ is convergent;

(b) and if $\lim_{n \to \infty} n\left(1 - \frac{x_{n+1}}{x_n}\right) < 1$, then $\sum x_n$ is divergent.

Show that we do not have any conclusion when

$$\lim_{n \to \infty} n\left(1 - \frac{x_{n+1}}{x_n}\right) = 1.$$

CHAPTER 8. SERIES   165

Problem 8.22 Consider the positive series $\sum x_n$ with $x_n > 0$ such that there exist $p > 0$ and $q > 1$ such that the sequence
$$\left\{ n^q \left(1 - \frac{x_{n+1}}{x_n} - \frac{p}{n} \right) \right\}$$
is bounded. Show that

(a) if $p \leq 1$, then $\sum x_n$ is divergent;

(b) and if $p > 1$, then $\sum x_n$ is convergent.

This is known as Raabe–Duhamel's rule.

Problem 8.23 Discuss the convergence or divergence of $\sum x_n$ where
$$x_n = \frac{1 \cdot 3 \cdots (2n-1)}{2 \cdot 4 \cdots (2n)}.$$

Solutions

Solution 8.1

Let $\sum a_k$ be a convergent series, then the sequence $\{s_n\}$ of partial sums is convergent and hence (s_n) is a Cauchy sequence. Thus for each $\varepsilon > 0$, there exists an $N \in \mathbb{N}$ such that

$$|s_n - s_m| < \varepsilon \qquad \text{for all} \quad n, m \geq N.$$

In particular,

$$|s_{n+1} - s_n| = \left|\sum_{k=0}^{n+1} a_k - \sum_{k=0}^{n} a_k\right| = |a_n| < \varepsilon \qquad \text{for all} \quad n \geq N,$$

that is, $\{a_n\}$ is a null sequence. Note also that if $\{s_n\}$ is convergent, then $\{s_{n+1}\}$ is also convergent and converges to the same limit. Hence $\{s_{n+1} - s_n\}$ converges to 0, i.e., $\{a_n\}$ converges to 0.

Solution 8.2

Note that $s_1 = a_1 = 2$ and

$$a_k = s_k - s_{k-1} = \frac{k+1}{k} - \frac{k}{k-1} = \frac{(k+1)(k-1) - k^2}{k(k-1)} = \frac{-1}{k(k+1)} \qquad \forall k > 1,$$

therefore the series is

$$2 - \sum_{k=2}^{\infty} \frac{1}{k(k+1)}$$

and the sum of the series is

$$s = \lim_{n \to \infty} s_n = \lim_{n \to \infty} \frac{n+1}{n} = 1.$$

Solution 8.3

Because the terms in the sum are all positive, the sequence of partial sums given by

$$s_n = 1 + \frac{1}{4} + \frac{1}{9} + \cdots + \frac{1}{n^2}$$

is increasing. To find an upper bound for s_n, observe

$$\begin{aligned}
s_n &= 1 + \frac{1}{2 \cdot 2} + \frac{1}{3 \cdot 3} + \cdots + \frac{1}{n \cdot n} \\
&< 1 + \frac{1}{2 \cdot 1} + \frac{1}{3 \cdot 2} + \cdots + \frac{1}{n \cdot (n-1)} \\
&= 1 + \left(1 - \frac{1}{2}\right) + \left(\frac{1}{2} - \frac{1}{3 \cdot 2}\right) + \cdots + \left(\frac{1}{n-1} - \frac{1}{n}\right) \\
&= 1 + 1 - \frac{1}{n} \\
&< 2.
\end{aligned}$$

Thus 2 is an upper bound for the sequence of partial sums, so by the Monotone Convergence Theorem, $\sum_{k=1}^{\infty} \frac{1}{k^2}$ converges to a limit less than 2.

Next, we claim that $\sum_{k=1}^{\infty} \frac{1}{k^2} = \frac{\pi^2}{6}$. This can be shown by using the well-known clever trick of evaluating the double integral

$$I = \int_0^1 \int_0^1 \frac{1}{1-xy} dx dy$$

and we evaluate I in two different ways. First notice

$$\frac{1}{1-xy} = \sum_{k=0}^{\infty} (xy)^k,$$

therefore

$$\begin{aligned} I &= \int_0^1 \int_0^1 \sum_{k=0}^{\infty} (xy)^k dx dy \\ &= \sum_{k=0}^{\infty} \int_0^1 \int_0^1 (xy)^k dx dy \\ &= \sum_{k=0}^{\infty} \left(\int_0^1 x^k dx \right) \left(\int_0^1 y^k dy \right) \\ &= \sum_{k=0}^{\infty} \frac{1}{(k+1)^2} \\ &= \sum_{k=1}^{\infty} \frac{1}{k^2}. \end{aligned}$$

The second way to evaluate I comes from a change of variables. Let

$$u = \frac{x+y}{2} \quad \text{and} \quad v = \frac{y-x}{2}$$

or equivalently

$$x = u - v \quad \text{and} \quad y = u + v.$$

Given this transformation,

$$\frac{1}{1-xy} = \frac{1}{1-(u^2+v^2)}$$

and using the change of variables formula we obtain

$$I = \int\int f(x,y) dx dy = \int\int f(x(u,v), y(u,v)) \left| \frac{d(x,y)}{d(u,v)} \right| du dv$$

where

$$\left| \frac{d(x,y)}{d(u,v)} \right| = 2.$$

Since the function to be integrated and the domain in the uv-plane are symmetric with respect to the u-axis, we can split the integral into two parts as such,

$$\begin{aligned} I & = 4\int_0^{1/2}\left(\int_0^u \frac{dv}{1-u^2+v^2}\right)du + 4\int_{1/2}^1\left(\int_0^{1-u}\frac{dv}{1-u^2+v^2}\right)du \\ & = 4\int_0^{1/2}\frac{1}{\sqrt{1-u^2}}\arctan\left(\frac{u}{\sqrt{1-u^2}}\right)du + 4\int_{1/2}^1\frac{1}{\sqrt{1-u^2}}\arctan\left(\frac{1-u}{\sqrt{1-u^2}}\right)du. \end{aligned}$$

Now, observe that if we set

$$k(u) = \arctan\left(\frac{u}{\sqrt{1-u^2}}\right) \quad \text{and} \quad h(u) = \arctan\left(\frac{1-u}{\sqrt{1-u^2}}\right),$$

then we obtain the derivatives

$$k'(u) = \frac{1}{\sqrt{u^2}} \quad \text{and} \quad h'(u) = -\frac{1}{2}\frac{1-u}{\sqrt{1-u^2}}.$$

This yields

$$\begin{aligned} I & = 4\int_0^{1/2} k'(u)k(u)du + 4\int_{1/2}^1 -2h'(u)h(u)du \\ & = 2\,(k(u))^2\big|_0^{1/2} - 4\,(h(u))^2\big|_{1/2}^1 \\ & = 2(k(1/2))^2 - 2(k(0))^2 - -4(h(1))^2 + 4(h(1/2))^2 \\ & = 2\left(\frac{\pi}{6}\right)^2 - 0 + 0 + 4\left(\frac{\pi}{6}\right)^2 \\ & = \left(\frac{\pi}{6}\right)^2. \end{aligned}$$

Solution 8.4

Again we have an increasing sequence of partial sums

$$s_k = 1 + \frac{1}{2} + \frac{1}{3} + \cdots + \frac{1}{k}.$$

Notice that

$$s_4 = 1 + \frac{1}{2} + \frac{1}{3} + \frac{1}{4} > 1 + \frac{1}{2} + \left(\frac{1}{4} + \frac{1}{4}\right) = 2.$$

A similar calculation yields

$$s_8 > 2\frac{1}{2},$$

therefore, in general we have

$$\begin{aligned} s_{2^k} & = 1 + \frac{1}{2} + \left(\frac{1}{3} + \frac{1}{4}\right) + \left(\frac{1}{5} + \cdots \frac{1}{8}\right) + \cdots + \left(\frac{1}{2^{k-1}} + \cdots \frac{1}{2^k}\right) \\ & > 1 + \frac{1}{2} + \left(\frac{1}{4} + \frac{1}{4}\right) + \left(\frac{1}{8} + \cdots \frac{1}{8}\right) + \cdots + \left(\frac{1}{2^k} + \cdots \frac{1}{2^k}\right) \\ & = 1 + \frac{1}{2} + 2\left(\frac{1}{4}\right) + 4\left(\frac{1}{8}\right) + \cdots + 2^{k-1}\left(\frac{1}{2^k}\right) \\ & = 1 + \frac{1}{2} + \frac{1}{2} + \frac{1}{2} + \cdots + \frac{1}{2} \\ & = 1 + k\left(\frac{1}{2}\right) \end{aligned}$$

CHAPTER 8. SERIES

which demonstrates that $\{s_{2^k}\}$ is unbounded. Despite the slow pace, the sequence of partial sums for $\sum_{k=1}^{\infty} \frac{1}{k}$ eventually surpasses every positive real number. The harmonic series diverges.

Solution 8.5

Assume $p > 0$. Otherwise we have the harmonic series which is divergent. The function $f(x) = \frac{1}{x \ln^p(x)}$ is decreasing on $[2, \infty)$. Then the integral test implies that $\sum_{n=2}^{\infty} \frac{1}{n \ln^p(n)}$ is convergent iff $\int_2^{\infty} f(x) dx$ is convergent. But this integral is the Bertrand improper integral. Using Problem 7.21, we know that $\int_2^{\infty} f(x) dx$ is convergent iff $p > 1$. Hence $\sum_{n=2}^{\infty} \frac{1}{n \ln^p(n)}$ is convergent iff $p > 1$. In particular, the series $\sum_{n=2}^{\infty} \frac{1}{n \ln(n)}$ is divergent.

Solution 8.6

a) Notice that
$$\frac{1}{k^2 + k} = \frac{1}{k(k+1)} = \frac{1}{k} + \frac{-1}{k+1}.$$

Therefore, the nth partial sum
$$\begin{aligned}
s_n &= \sum_{k=1}^{\infty} \frac{1}{k} - \frac{1}{k+1} \\
&= \left(1 - \frac{1}{2}\right) + \left(\frac{1}{2} - \frac{1}{3}\right) + \cdots + \left(\frac{1}{n} - \frac{1}{n+1}\right) \\
&= 1 + \left(-\frac{1}{2} + \frac{1}{2}\right) + \left(-\frac{1}{3} + \frac{1}{3}\right) + \cdots + \left(-\frac{1}{n} + \frac{1}{n}\right) - \frac{1}{n+1} \\
&= 1 - \frac{1}{n+1}.
\end{aligned}$$

The limit of the sequence of partial sums is
$$\lim_{n \to \infty} s_n = \lim \left(1 - \frac{1}{n+1}\right) = 1$$

so the series converges with sum 1.

b) Using the properties of logarithms, we see that
$$\sum_{k=1}^{\infty} \frac{\ln\left(\frac{k^{k+1}}{(k+1)^k}\right)}{k(k+1)} = \sum_{k=1}^{\infty} \frac{(k+1) \ln k - k \ln(k+1)}{k(k+1)}$$

and therefore

$$s_n = \sum_{k=1}^{n} \frac{\ln k}{k} - \frac{\ln(k+1)}{k+1}$$
$$= \left(\frac{\ln 1}{1} - \frac{\ln 2}{2}\right) + \left(\frac{\ln 2}{2} - \frac{\ln 3}{3}\right) + \cdots + \left(\frac{\ln n}{n} - \frac{\ln(n+1)}{n+1}\right)$$
$$= -\frac{\ln(n+1)}{n+1},$$

which implies

$$\lim_{n \to \infty} s_n = \lim \left(-\frac{\ln(n+1)}{n+1}\right) = 0.$$

c) We utilize the proof of Problem 8.3 to claim that

$$\sum_{k=1}^{\infty} \frac{1}{k^2} = \frac{\pi^2}{6}.$$

Next we observe that

$$s_{2n} = \sum_{k=1}^{2n} \frac{1}{k^2} = \sum_{k=1}^{n} \frac{1}{(2k)^2} + \sum_{k=1}^{n} \frac{1}{(2k-1)^2}.$$

We now take the limit of s_{2n} and find that

$$\lim_{n \to \infty} s_{2n} = \frac{1}{4} \sum_{k=1}^{\infty} \frac{1}{(2k)^2} + \sum_{k=1}^{\infty} \frac{1}{(2k-1)^2}.$$

Therefore, it is the case that

$$\sum_{k=1}^{\infty} \frac{1}{(2k-1)^2} = \frac{3}{4} \sum_{k=1}^{\infty} \frac{1}{(2k)^2} = \frac{3}{4} \frac{\pi^2}{6} = \frac{\pi^2}{8}.$$

Solution 8.7

Note that if a series $\sum x_n$ is convergent, then we must have $\lim_{n \to \infty} x_n = 0$. Obviously this will imply that $\left\{\frac{1}{x_n}\right\}$ is divergent. Hence $\sum \frac{1}{x_n}$ is divergent. For the converse, take $x_n = \sqrt{n}$, then both $\sum x_n$ and $\sum \frac{1}{x_n}$ are divergent. So the converse is false.

Solution 8.8

Since the series $\sum x_n$ is convergent, we must have $\lim_{n \to \infty} x_n = 0$. So there exists $N \geq 1$ such that $x_n < 1$, for $n \geq N$. Hence $x_n^2 < x_n$ for $n \geq N$. Since $\sum_{n \geq N} x_n$ is convergent, $\sum_{n \geq N} x_n^2$ is also

CHAPTER 8. SERIES

convergent, which implies $\sum_{n \geq 1} x_n^2$ is convergent. For the series $\sum \sqrt{x_n x_{n+1}}$ use the algebraic identity $\sqrt{\alpha \beta} \leq \dfrac{\alpha + \beta}{2}$ for any α and β positive numbers, to get

$$\sqrt{x_n x_{n+1}} \leq \frac{x_n + x_{n+1}}{2}$$

for $n \geq 1$. Since the series $\sum x_n$ and $\sum x_{n+1}$ are convergent, we conclude that $\sum \sqrt{x_n x_{n+1}}$ is also convergent.

Solution 8.9

Indeed, we know that for any positive real numbers α and β, we have $\alpha\beta \leq \dfrac{\alpha^2 + \beta^2}{2}$. So

$$\frac{\sqrt{x_n}}{n} \leq \frac{x_n + \dfrac{1}{n^2}}{2} = \frac{x_n}{2} + \frac{1}{2n^2}$$

for $n \geq 1$. Since the series $\sum \dfrac{x_n}{2}$ and $\sum \dfrac{1}{2n^2}$ are convergent, we conclude that the original series $\sum \dfrac{\sqrt{x_n}}{n}$ is convergent.

Solution 8.10

It is well known from the alternating series test that the series $\sum \dfrac{(-1)^n}{\sqrt{n}}$ is convergent. Let us focus on the series $\sum \dfrac{(-1)^n}{\sqrt{n} + (-1)^n}$. We have

$$\frac{(-1)^n}{\sqrt{n} + (-1)^n} - \frac{(-1)^n}{\sqrt{n}} + \frac{1}{n} = (-1)^n \left[\frac{-(-1)^n}{\sqrt{n}\left(\sqrt{n} + (-1)^n\right)} \right] + \frac{1}{n}$$

or

$$\frac{(-1)^n}{\sqrt{n} + (-1)^n} - \frac{(-1)^n}{\sqrt{n}} + \frac{1}{n} = \frac{-1}{n + (-1)^n \sqrt{n}} + \frac{1}{n} = \frac{(-1)^n \sqrt{n}}{n\left(n + (-1)^n \sqrt{n}\right)}.$$

Using the inequality $\left||a| - |b|\right| \leq |a - b|$ for any real numbers a and b, we get $n^2 - n\sqrt{n} \leq |n^2 - (-1)^n n\sqrt{n}|$, for any $n \geq 2$. This will imply

$$\left| \frac{(-1)^n}{\sqrt{n} + (-1)^n} - \frac{(-1)^n}{\sqrt{n}} + \frac{1}{n} \right| \leq \frac{\sqrt{n}}{n^2 - n\sqrt{n}}$$

for any $n \geq 2$. Note that for any $n \geq 4$, we have $2\sqrt{n} \leq n$ or $\sqrt{n} \leq n - \sqrt{n}$. Hence

$$\left| \frac{(-1)^n}{\sqrt{n} + (-1)^n} - \frac{(-1)^n}{\sqrt{n}} + \frac{1}{n} \right| \leq \frac{\sqrt{n}}{n\sqrt{n}}$$

for any $n \geq 4$. Since the series $\sum \frac{1}{n^{1.5}}$ is convergent, from the basic *comparison theorem* we get that $\sum \left| \frac{(-1)^n}{\sqrt{n}+(-1)^n} - \frac{(-1)^n}{\sqrt{n}} + \frac{1}{n} \right|$ is convergent. Hence the series $\sum \left(\frac{(-1)^n}{\sqrt{n}+(-1)^n} - \frac{(-1)^n}{\sqrt{n}} + \frac{1}{n} \right)$ is absolutely convergent which implies that it is also convergent. Since

$$\frac{(-1)^n}{\sqrt{n}+(-1)^n} = \frac{(-1)^n}{\sqrt{n}} - \frac{1}{n} + \left(\frac{(-1)^n}{\sqrt{n}+(-1)^n} - \frac{(-1)^n}{\sqrt{n}} + \frac{1}{n} \right)$$

and the series $\sum \frac{(-1)^n}{\sqrt{n}}$ is convergent while $\sum \frac{1}{n}$ is divergent, we can deduce that $\sum \frac{(-1)^n}{\sqrt{n}+(-1)^n}$ is divergent. Because

$$\lim_{n \to \infty} \frac{\frac{(-1)^n}{\sqrt{n}+(-1)^n}}{\frac{(-1)^n}{\sqrt{n}}} = \lim_{n \to \infty} \frac{\sqrt{n}}{\sqrt{n}+(-1)^n} = \lim_{n \to \infty} \frac{1}{1+\frac{(-1)^n}{\sqrt{n}}} = 1$$

we conclude that the limit test is not valid for nonpositive series.

Solution 8.11

We have

$$\int_1^a e^{-x^n} dx \leq \int_1^\infty e^{-x^n} dx = x_n, \quad n = 1, \ldots,$$

for any $a \geq 1$. Set $\varepsilon_n = \frac{1}{n \ln(n)}$, for $n \geq 2$. Then we have

$$\int_1^{1+\varepsilon_n} e^{-x^n} dx \leq \int_1^\infty e^{-x^n} dx = x_n, \quad n = 2, \ldots.$$

Since e^{-x^n} is decreasing on $[0, \infty)$, we have

$$\varepsilon_n e^{-(1+\varepsilon_n)^n} \leq \int_1^{1+\varepsilon_n} e^{-x^n} dx \leq x_n, \quad n = 2, \ldots.$$

On the other hand, we have

$$\lim_{n \to \infty} (1+\varepsilon_n)^n = \lim_{n \to \infty} e^{n \ln(1+\varepsilon_n)} = e^0 = 1.$$

Since $\sum \varepsilon_n$ is divergent, because it is a Bertrand series (see Problem 8.5), the series $\sum \varepsilon_n e^{-(1+\varepsilon_n)^n}$ is also divergent because of the limit test and both series are positive. The basic comparison test will force $\sum x_n$ to be divergent.

Solution 8.12

Indeed, let $n \geq 2$ and $N > n$. Then

$$\sum_{k=n}^N \varepsilon_k x_k = \sum_{k=n}^N \varepsilon_k (s_k - s_{k-1}) = \sum_{k=n}^N \varepsilon_k s_k - \sum_{k=n}^N \varepsilon_k s_{k-1}.$$

CHAPTER 8. SERIES

But
$$\sum_{k=n}^{N} \varepsilon_k s_{k-1} = \sum_{k=n-1}^{N-1} \varepsilon_{k+1} s_k.$$

Hence
$$\sum_{k=n}^{N} \varepsilon_k x_k = \sum_{k=n}^{N-1} (\varepsilon_k - \varepsilon_{k+1}) s_k + \varepsilon_N s_N - \varepsilon_n s_{n-1}.$$

Let $\varepsilon > 0$. Then since $\{\varepsilon_n\}$ goes to 0 and $\{s_n\}$ is bounded, there exists $n_0 \geq 2$ such that for any $n, N \geq n_0$ we have $|\varepsilon_N s_N - \varepsilon_n s_{n-1}| < \varepsilon/2$. Also since $\sum |\varepsilon_{n+1} - \varepsilon_n|$ is convergent and $\{s_n\}$ is bounded, the basic comparison test implies that $\sum |(\varepsilon_{n+1} - \varepsilon_n) s_n|$ is also convergent. Therefore there exists $n_1 \geq 2$ such that for any $n, N \geq n_1$ we have

$$\sum_{k=n}^{N-1} |(\varepsilon_k - \varepsilon_{k+1}) s_k| < \frac{\varepsilon}{2}.$$

Let $n_2 > \max\{n_0, n_1\}$. Then for any $n, N \geq n_2$, we have

$$\left| \sum_{k=n}^{N} \varepsilon_k x_k \right| \leq \sum_{k=n}^{N-1} |(\varepsilon_k - \varepsilon_{k+1}) s_k| + |\varepsilon_N s_N - \varepsilon_n s_{n-1}| < \varepsilon.$$

The Cauchy criterion for series will imply that the series $\sum \varepsilon_n x_n$ is convergent as claimed.

Solution 8.13

We know that $\sum_{n \in \mathbb{N}} \frac{1}{n!}$ is convergent and its sum is e, i.e.,

$$\sum_{n=0}^{\infty} \frac{1}{n!} = e.$$

This proves one side of the inequalities. Set

$$x_n = 1 + \frac{1}{2!} + \cdots + \frac{1}{n(n!)}.$$

Then clearly we have $\lim_{n \to \infty} x_n = e$. Let u show that in fact $\{x_n\}$ is decreasing. Indeed we have

$$x_{n+1} - x_n = \frac{1}{(n+1)!} + \frac{1}{(n+1)(n+1)!} - \frac{1}{n(n)!} = \frac{1}{n!} \left(\frac{1}{n+1} + \frac{1}{(n+1)^2} - \frac{1}{n} \right).$$

Since
$$\frac{1}{n+1} + \frac{1}{(n+1)^2} - \frac{1}{n} = \frac{n(n+1) + n - (n+1)^2}{n(n+1)^2} = \frac{-1}{n(n+1)^2},$$

then $x_{n+1} - x_n < 0$, for $n \geq 1$. So $\{x_n\}$ is decreasing which implies $e \leq x_n$, for $n \geq 1$. Therefore we have

$$1 + \frac{1}{2!} + \cdots + \frac{1}{n!} \leq e \leq 1 + \frac{1}{2!} + \cdots + \frac{1}{n(n!)},$$

for $n \geq 1$. These inequalities will imply $m\pi \leq n!\pi e \leq m\pi + \pi/n$, where

$$m = n!\left(1 + \frac{1}{2!} + \cdots + \frac{1}{n!}\right) \in \mathbb{N}.$$

Hence $\sin(n!\pi e) = (-1)^m \sin(\pi \varepsilon_n)$, where $0 \leq \varepsilon_n \leq 1/n$, for $n \geq 1$. Note that $(-1)^m = (-1)^{n+1}$. Indeed we have

$$m = n!\left(1 + \frac{1}{2!} + \cdots + \frac{1}{n!}\right) = n(n-1)k + n + 1$$

where $k \in \mathbb{N}$. Since $n(n-1)$ is even, we get $(-1)^m = (-1)^{n+1}$. Also note that

$$\varepsilon_n = n!\left(e - 1 - \frac{1}{2!} - \cdots - \frac{1}{n!}\right) = n! \sum_{k=n+1}^{\infty} \frac{1}{k!}.$$

Let us prove that $\{\varepsilon_n\}$ is decreasing. Indeed we have

$$\varepsilon_{n+1} - \varepsilon_n = (n+1)! \sum_{k=n+2}^{\infty} \frac{1}{k!} - n! \sum_{k=n+1}^{\infty} \frac{1}{k!} = \big((n+1)! - n!\big) \sum_{k=n+2}^{\infty} \frac{1}{k!} - \frac{n!}{(n+1)!}$$

which implies

$$\varepsilon_{n+1} - \varepsilon_n = n! \sum_{k=n+2}^{\infty} \frac{1}{k!} - \frac{n!}{(n+1)!} = n!\left(\sum_{k=n+2}^{\infty} \frac{1}{k!} - \frac{1}{(n+1)!}\right).$$

On the other hand, we have

$$(n+1)! \sum_{k=n+2}^{\infty} \frac{1}{k!} = \frac{1}{n+2} + \frac{1}{(n+2)(n+3)} + \frac{1}{(n+2)(n+3)(n+4)} + \cdots.$$

But

$$\frac{1}{(n+2)(n+3)\cdots(n+k)} \leq \frac{1}{(n+k-1)(n+k)} = \frac{1}{n+k-1} - \frac{1}{n+k}$$

which implies

$$(n+1)! \sum_{k=n+2}^{\infty} \frac{1}{k!} \leq \frac{2}{n+2} < 1$$

for $n \geq 1$. Hence $\varepsilon_{n+1} - \varepsilon_n < 0$, for $n \geq 1$, i.e., $\{\varepsilon_n\}$ is a decreasing sequence of positive numbers. And since $\varepsilon_n \leq 1/n$, we get $\lim_{n \to \infty} \varepsilon_n = 0$. Putting everything together and using the alternating series test we conclude that $\sum_{n \geq 1} \sin(\pi e n!)$ is convergent.

$\boxed{\text{Solution 8.14}}$

Since $\left\{\dfrac{1}{pn+1}\right\}$ is decreasing and goes to 0, the alternating series test implies that $\sum x_n$ is convergent, for any $p > 0$. Let us find its sum. Let $N \in \mathbb{N}$. Then we have

$$\sum_{n=0}^{N} \frac{(-1)^n}{pn+1} = \sum_{n=0}^{N} (-1)^n \int_0^1 t^{pn} dt = \int_0^1 \left(\sum_{n=0}^{N} (-1)^n t^{pn}\right) dt = \int_0^1 \frac{1 - (-t^p)^{N+1}}{1 + t^p} dt.$$

CHAPTER 8. SERIES

Hence
$$\left| \sum_{n=0}^{N} \frac{(-1)^n}{p\,n+1} - \int_0^1 \frac{1}{1+t^p}dt \right| \leq \int_0^1 \frac{t^{p(N+1)}}{1+t^p}dt = \frac{1}{p(N+1)+1}.$$

If we let $N \to \infty$, we get the desired relation
$$\sum_{n=0}^{\infty} x_n = \int_0^1 \frac{dt}{1+t^p}.$$

In particular, if we let $p = 1$, then we have
$$\sum_{n=0}^{\infty} \frac{(-1)^n}{n+1} = \int_0^1 \frac{dt}{1+t} = \Big[\ln(t+1)\Big]_0^1 = \ln 2.$$

If we let $p = 2$, we get
$$\sum_{n=0}^{\infty} \frac{(-1)^n}{2n+1} = \int_0^1 \frac{dt}{1+t^2} = \Big[\arctan(t)\Big]_0^1 = \frac{\pi}{4}.$$

And finally if we let $p = 3$, we get
$$\sum_{n=0}^{\infty} \frac{(-1)^n}{3n+1} = \int_0^1 \frac{dt}{1+t^3}.$$

To perform this integral we will make use of the partial decomposition technique to get
$$\frac{1}{1+t^3} = \frac{1}{3}\left(\frac{1}{1+t} - \frac{t-2}{t^2-t+1}\right) = \frac{1}{3}\left(\frac{1}{1+t} - \frac{1}{2}\frac{2t-1}{t^2-t+1} + \frac{3}{2}\frac{1}{(t-1/2)^2+3/4}\right)$$

which implies
$$\int \frac{dt}{1+t^3} = \frac{1}{3}\left(\ln(1+t) - \frac{1}{2}\ln(t^2-t+1) + \sqrt{3}\arctan\left(\frac{2t-1}{\sqrt{3}}\right)\right) + C.$$

Since
$$\int_0^1 \frac{dt}{1+t^3} = \frac{1}{3}\left(\ln(2) + \frac{\pi}{\sqrt{3}}\right),$$

we get
$$\sum_{n=0}^{\infty} \frac{(-1)^n}{3n+1} = \frac{1}{3}\left(\ln(2) + \frac{\pi}{\sqrt{3}}\right).$$

Solution 8.15

We will make use of Abel's theorem (see Problem 8.12). Recall Euler's formula
$$e^{i\theta} = \cos(\theta) + i\sin(\theta).$$

We have
$$\sum_{n=1}^{N} e^{in\theta} = e^{i\theta}\frac{1-e^{Ni\theta}}{1-e^{i\theta}} = e^{i(N+1)\theta/2}\frac{\sin(N\theta/2)}{\sin(\theta/2)}$$

for any $N \geq n \geq 1$. Note that we made use of $e^{i\theta} \neq 1$. Hence

$$\left|\sum_{n=1}^{N} e^{in\theta}\right| = \left|\frac{\sin(N\theta/2)}{\sin(\theta/2)}\right| \leq \frac{1}{|\sin(\theta/2)|}.$$

Since $\{1/n\}$ decreases and goes to 0, then by Abel's theorem, the series

$$\sum_{n\geq 1} \frac{\cos(n\theta)}{n} \quad \text{and} \quad \sum_{n\geq 1} \frac{\sin(n\theta)}{n}$$

are convergent.

Solution 8.16

Since $\sum y_r$ is a rearrangement of $\sum x_r$, there is a bijection $f : \mathbb{N} \to \mathbb{N}$ such that $y_r = x_{f(r)}$; and $x_r = y_{f^{-1}(r)}$.

Let $\varepsilon > 0$. Since $\sum x_r$ is absolutely convergent, it follows from the General Principle of Convergence that there exists N such that

$$n \geq m \geq N \quad \Longrightarrow \quad \sum_{r=m}^{n} |x_r| < \frac{1}{2}\varepsilon.$$

In fact, if S is any finite subset of $\{r \in \mathbb{N} : r > N\}$, then

(8.1) $$\sum_{r\in S} |x_r| < \frac{1}{2}\varepsilon.$$

Next, we show that

(8.2) $$m \geq N \quad \Longrightarrow \quad \left|s - \sum_{r=1}^{m} x_r\right| \leq \frac{1}{2}\varepsilon.$$

The reason for this is that

$$\left|s - \sum_{r=1}^{m} x_r\right| = \lim_{n\to\infty} \left|\sum_{r=1}^{n} x_r - \sum_{r=1}^{m} x_r\right|$$

$$= \lim_{n\to\infty} \left|\sum_{r=m+1}^{n} x_r\right| \leq \lim_{n\to\infty} \sum_{r=m+1}^{n} |x_r| \leq \frac{1}{2}\varepsilon$$

by (8.1).

Now let $M = \max\{f^{-1}(1), \ldots, f^{-1}(N)\}$, and note that $M \geq N$. Since $x_r = y_{f^{-1}(r)}$, if follows that $\{y_1, \ldots, y_M\} \supset \{x_1, \ldots, x_N\}$. So, if $n \geq M$, then

$$\{y_1, \ldots, y_n\} = \{x_1, \ldots, x_N\} \cup \{x_r : r \in S_n\},$$

for some finite subset S_n of $\{r \in \mathbb{N}\}$. Thus

$$n \geq M \quad \Longrightarrow \quad \left|s - \sum_{r=1}^{n} y_r\right| \leq \left|s - \sum_{r=1}^{N} x_r\right| + \sum_{r\in S_n} |x_r|$$

$$< \frac{1}{2}\varepsilon + \frac{1}{2}\varepsilon = \varepsilon$$

CHAPTER 8. SERIES

by (8.2) and (8.1). This shows that $\sum y_r$ converges to s.

Solution 8.17

We show that the series

(8.3) $\quad x_0y_0 + x_0y_1 + x_1y_1 + x_1y_0 + x_0y_2 + x_1y_2 + x_2y_2 + x_2y_1 + x_2y_0 + \cdots$

converges absolutely. In fact, the sum of the first $(n+1)^2$ terms of the series

(8.4) $\quad |x_0y_0| + |x_0y_1| + |x_1y_1| + |x_1y_0| + |x_0y_2| + |x_1y_2| + |x_2y_2| + |x_2y_1| + |x_2y_0| + \cdots$

is $\sum_{k=0}^{n} |x_k| \cdot \sum_{k=0}^{n} |y_k|$, which converges to $\sum_{k=0}^{\infty} |x_k| \cdot \sum_{k=0}^{\infty} |y_k|$.

Hence the sequence of partial sums of (8.4) has a convergent subsequence. But all the terms of the series are positive, so the sequence of all partial sums is increasing, and bounded above by $\sum_{k=0}^{\infty} |x_k| \cdot \sum_{k=0}^{\infty} |y_k|$, so it also converges, to the same limit.

Thus (8.3) converges absolutely.

The same argument as above, considering sums of the first $(n+1)^2$ terms of the series (8.3), shows that the sum of this series is $\sum x_n \cdot \sum y_n$.

But
$$\sum z_n = x_0y_0 + (x_0y_1 + x_1y_0) + (x_0y_2 + x_1y_1 + x_2y_0) + \cdots$$

is a rearrangement of (8.3), so by the Rearrangement Theorem it converges to the same limit.
Note: You might like to think about the above proof by considering an "infinite matrix" in which the (i,j) term is x_iy_i:

$$\begin{array}{ccccc} x_0y_0 & x_0y_1 & x_0y_2 & x_0y_3 & \cdots \\ x_1y_0 & x_1y_1 & x_1y_2 & x_1y_3 & \cdots \\ x_2y_0 & x_2y_1 & x_2y_2 & x_2y_3 & \cdots \\ x_3y_0 & x_3y_1 & x_3y_2 & x_3y_3 & \cdots \\ \vdots & \vdots & \vdots & \vdots & \ddots \end{array}$$

Solution 8.18

Note that since all the terms are positive, we have
$$\frac{x_{n+1}}{x_n} \frac{x_n}{x_{n-1}} \cdots \frac{x_{N+1}}{x_N} \leq \frac{y_{n+1}}{y_n} \frac{y_n}{y_{n-1}} \cdots \frac{y_{N+1}}{y_N}$$

which implies
$$\frac{x_{n+1}}{x_N} \leq \frac{y_{n+1}}{y_N}$$

for any $n \geq N$. Hence
$$\frac{y_N}{x_N} x_n \leq y_n$$

for any $n \geq N$. This will imply our desired conclusion from the basic comparison test.

Solution 8.19

Set $y_n = \dfrac{1}{(n-1)^p}$. Then
$$\frac{y_{n+1}}{y_n} = \left(\frac{n-1}{n}\right)^p = \left(1 - \frac{1}{n}\right)^p.$$

Using the inequality $(1-x)^p \geq 1 - p\,x$ for any $x \in [0,1]$, we get

$$\frac{y_{n+1}}{y_n} \geq 1 - p\frac{1}{n} \geq \frac{x_{n+1}}{x_n}$$

for any $n \geq N$, because of our assumption on $\{x_n\}$. Since $p > 1$, the series $\sum_{n \geq N} y_n$ is convergent. From the previous problem we can deduce that $\sum_{n \geq N} x_n$ is convergent, which implies that $\sum x_n$ is convergent.

Solution 8.20

Set $y_n = \dfrac{1}{(n-1)}$. Then

$$\frac{y_{n+1}}{y_n} = \frac{n-1}{n} = 1 - \frac{1}{n}.$$

Hence

$$\frac{y_{n+1}}{y_n} \leq \frac{x_{n+1}}{x_n}$$

for any $n \geq N$, because of our assumption on $\{x_n\}$. Since the series $\sum_{n \geq N} y_n$ is divergent, the previous problem implies that $\sum_{n \geq N} x_n$ is divergent which means that $\sum x_n$ is divergent.

Solution 8.21

For the proof of (a). Let $p \in \mathbb{R}$ such that

$$1 < p < \lim_{n \to \infty} n\left(1 - \frac{x_{n+1}}{x_n}\right).$$

Then there exists $N \geq 1$ such that

$$n\left(1 - \frac{x_{n+1}}{x_n}\right) \geq p.$$

Since $p > 1$, Problem 8.19 implies that $\sum x_n$ is convergent. Similar ideas will imply the conclusion of (b). Finally, if we take $x_n = 1/n$, then

$$\lim_{n \to \infty} n\left(1 - \frac{x_{n+1}}{x_n}\right) = \lim_{n \to \infty} n\left(1 - \frac{n}{n+1}\right) = \lim_{n \to \infty} \frac{n}{n+1} = 1.$$

And if we take $x_n = 1/n \ln^2(n)$, then

$$\lim_{n \to \infty} n\left(1 - \frac{x_{n+1}}{x_n}\right) = 1.$$

Indeed, we have

$$n\left(1 - \frac{x_{n+1}}{x_n}\right) = n\left(1 - \frac{n \ln^2(n)}{(n+1)\ln^2(n+1)}\right) = \frac{n}{n+1}\left(\frac{(n+1)\ln^2(n+1) - n\ln^2(n)}{\ln^2(n+1)}\right),$$

but

$$\frac{(n+1)\ln^2(n+1) - n\ln^2(n)}{\ln^2(n+1)} = \frac{n\left(\ln^2(n+1) - \ln^2(n)\right)}{\ln^2(n+1)} + 1 = n\ln\left(1+\frac{1}{n}\right)\frac{\ln(n+1) + \ln(n)}{\ln^2(n+1)} + 1.$$

Since

$$\lim_{n\to\infty} n\ln\left(1+\frac{1}{n}\right) = 1, \text{ and } \lim_{n\to\infty} \frac{\ln(n+1) + \ln(n)}{\ln^2(n+1)} = \lim_{n\to\infty} \frac{1}{\ln(n+1)}\left(1 + \frac{\ln(n)}{\ln(n+1)}\right) = 0,$$

we obtain

$$\lim_{n\to\infty} n\left(1 - \frac{x_{n+1}}{x_n}\right) = 1.$$

So for both sequences the above condition is satisfied but $\sum \frac{1}{n}$ is divergent while $\sum \frac{1}{n\ln^2(n)}$ is convergent (see Problem 8.5).

Solution 8.22

Our assumption implies the existence of $M > 0$ such that for any $n \geq 1$ we have

$$\left|1 - \frac{x_{n+1}}{x_n} - \frac{p}{n}\right| \leq \frac{M}{n^q}.$$

So

$$\left|n\left(1 - \frac{x_{n+1}}{x_n}\right) - p\right| \leq \frac{M}{n^{q-1}}.$$

Hence

$$\lim_{n\to\infty} n\left(1 - \frac{x_{n+1}}{x_n}\right) = p.$$

Problem 8.21 implies that if $p < 1$, then $\sum x_n$ is divergent, and if $p > 1$, then $\sum x_n$ is convergent. So assume $p = 1$ and let us prove that $\sum x_n$ is divergent. Indeed, set

$$\varepsilon_n = n^q\left(\frac{x_{n+1}}{x_n} - 1 + \frac{1}{n}\right),$$

then

$$\frac{x_{n+1}}{x_n} = 1 - \frac{1}{n} + \frac{\varepsilon_n}{n^q}.$$

Set $y_n = nx_n$, then

$$\frac{y_{n+1}}{y_n} = \frac{n+1}{n}\frac{x_{n+1}}{x_n} = \left(1 + \frac{1}{n}\right)\left(1 - \frac{1}{n} + \frac{\varepsilon_n}{n^q}\right)$$

which implies

$$\frac{y_{n+1}}{y_n} = 1 - \frac{1}{n^2} + \frac{\delta_n}{n^q}$$

where $\delta_n = (n+1)\varepsilon_n/n$. Since $\{\delta_n\}$ is bounded,

$$\lim_{n\to\infty} n^\alpha \ln\left(\frac{y_{n+1}}{y_n}\right) = 1,$$

where $\alpha = \min(2, q)$. Thus the series $\sum \ln\left(\dfrac{y_{n+1}}{y_n}\right)$ is convergent by the limit test which implies that $\lim_{n \to \infty} \ln(y_n) = l$ exists. Hence $\lim_{n \to \infty} y_n = e^l$, i.e.,

$$\lim_{n \to \infty} n x_n = e^l.$$

The limit test again implies that $\sum x_n$ is divergent since $\sum e^l/n$ is divergent.

Solution 8.23

Note that
$$x_n = \frac{(2n)!}{(2^n n!)^2} = \frac{(2n)!}{2^{2n}(n!)^2}.$$

Let us first remark that $\lim_{n \to \infty} x_n = 0$. Indeed, notice that $0 < x_n < 1$ for any $n \geq 1$. Since $x_{n+1} = \dfrac{2n+1}{2n+2} x_n$, we have $x_{n+1} < x_n$ for any $n \geq 1$, which implies that $\{x_n\}$ is decreasing. So $\lim_{n \to \infty} x_n = l$ exists. Define

$$y_n = \frac{2 \cdot 4 \cdots (2n)}{3 \cdot 5 \cdots (2n+1)}.$$

Since $4n^2 - 1 < 4n^2$ we get $\dfrac{2n-1}{2n} < \dfrac{2n}{2n+1}$, for any $n \geq 1$. This obviously implies $x_n < y_n$ for any $n \geq 1$. Since $x_n y_n = \dfrac{1}{2n+1}$ we deduce that $x_n^2 < x_n y_n = \dfrac{1}{2n+1}$ for any $n \geq 1$. Hence $\lim_{n \to \infty} x_n^2 = 0$ which implies $\lim_{n \to \infty} x_n = 0$. This conclusion may suggest that the series $\sum x_n$ is convergent. But since

$$\frac{x_{n+1}}{x_n} = \frac{2n+1}{2n+2},$$

we have

$$n\left(1 - \frac{x_{n+1}}{x_n}\right) = \frac{n}{2n+2}.$$

Hence $\lim_{n \to \infty} n\left(1 - \dfrac{x_{n+1}}{x_n}\right) = \dfrac{1}{2} < 1$, then $\sum x_n$ is divergent based on the previous problem. Note that the ratio test does not help since $\lim_{n \to \infty} \dfrac{x_{n+1}}{x_n} = 1$. In fact, the root test will also be not conclusive.

Chapter 9

Metric Spaces

Therefore, either the reality on which our space is based must form a discrete manifold or else the reason for the metric relationships must be sought for, externally, in the binding forces acting on it.

Bernhard Riemann (1826–1866)

- Let X be a set. A function
$$d : X \times X \to \mathbb{R}^+$$
is called a *metric* on X if the following hold:

  (M1) $d(x,y) = 0 \Leftrightarrow x = y$,
  (M2) $d(x,y) = d(y,x)$ for all $x, y \in X$ (Symmetry), and
  (M3) $d(x,y) \leq d(x,z) + d(z,y)$ for all $x, y, z \in X$ (Triangle Inequality).

  If d is a metric on X, then (X,d) is called a *metric space*.

- Let (X,d) be a metric space. A sequence $(x_n) \subseteq X$ *converges* to an element $x \in X$ if for all $\varepsilon > 0$ there exists an $N \in \mathbb{N}$ such that $d(x_n, x) < \varepsilon$ whenever $n \geq N$.

- A sequence in a metric space (X,d) is a *Cauchy sequence* if for all $\varepsilon > 0$ there exists an $N \in \mathbb{N}$ such that $d(x_n, x_m) < \varepsilon$ whenever $n, m \geq N$.

- A metric space (X,d) is called *complete* if every Cauchy sequence in X converges to an element of X.

Problem 9.1 Show that the following functions define a metric on $\mathbb{R}$.

(a) $d(x,y) = |x-y|$

(b) $d(x,y) = \sqrt{|x-y|}$

Problem 9.2 Show that the function
$$d(x,y) = \begin{cases} 0 & \text{if } x = y, \\ 1 & \text{if } x \neq y \end{cases}$$
is a metric on any nonempty set X. (This is called the *discrete metric* on X.)

Problem 9.3 Let $X = \mathbb{R}^2$ be the set of points in the plane. Show that for $\mathbf{x}, \mathbf{y} \in X$ with $\mathbf{x} = (x_1, x_2)$ and $\mathbf{y} = (y_1, y_2)$,
$$d(\mathbf{x},\mathbf{y}) = \sqrt{(x_1-y_1)^2 + (x_2-y_2)^2}$$
is a metric on X, also known as the Euclidean metric on $\mathbb{R}^2$.

Problem 9.4 Let $X = \mathbb{R}^2$. Show that for $\mathbf{x} = (x_1, x_2), \mathbf{y} = (y_1, y_2) \in X$, the function
$$d(\mathbf{x},\mathbf{y}) = \begin{cases} |x_1 - y_1| & \text{if } x_2 = y_2, \\ |x_1| + |x_2 - y_2| + |y_1| & \text{if } x_2 \neq y_2 \end{cases}$$
is a metric on X.

Problem 9.5 Let X be the set of continuous functions from $[a,b]$ to $\mathbb{R}$. For all $x, y \in X$ define $d(x,y)$ to be
$$d(x,y) = \max\{|x(t) - y(t)| : t \in [a,b]\}.$$
Show that (X, d) is a metric space.

Problem 9.6 Let X be the set of bounded functions from some set A to $\mathbb{R}$. For $x, y \in X$ define $d(x,y)$ by
$$d(x,y) = \sup\{|x(t) - y(t)| : t \in A\}.$$
Show that (X, d) is a metric space.

CHAPTER 9. METRIC SPACES

Problem 9.7 Let X be the set of continuous functions from $[a,b]$ into $\mathbb{R}$. For $x, y \in X$ define $d(x, y)$ by
$$d(x, y) = \int_a^b |x(t) - y(t)|\, dt.$$
Prove that (X, d) is a metric space.

Problem 9.8 Let (X, d) be a metric space. Show for all $x, y, z, w \in X$ we have
$$\left| d(x, z) - d(z, y) \right| \leq d(x, y)$$
and
$$\left| d(x, y) - d(z, w) \right| \leq d(x, z) + d(y, w).$$

Problem 9.9 Let (X, d) be a metric space. Suppose ρ is defined by
$$\rho(x, y) = \frac{d(x, y)}{1 + d(x, y)}.$$
Show that ρ is also a metric on X.
(Note that the new metric ρ is bounded because $\rho(x, y) < 1$ for all $x, y \in X$.)

Problem 9.10 Let (X, d) be a metric space. Let p be a point in X. The *SNCF* metric d_p is defined by
$$d_p(x, y) := \begin{cases} 0 & \text{if } x = y, \\ d(x, p) + d(p, y) & \text{otherwise.} \end{cases}$$
Show that d_p is a metric.

Problem 9.11 Let (X, d) be a metric space. The metric d is called an *ultrametric* if
$$d(x, y) \leq \max \left\{ d(x, z), d(y, z) \right\}$$
for any $x, y, z \in X$.

(i) Show that the Euclidean metric (see Problem 9.3) on $\mathbb{R}^n$, for $n \geq 2$, is not an ultrametric.

(ii) Let p be any positive prime number $p \geq 2$. For any nonzero $x \in \mathbb{Q}$, there exists a unique $n \in \mathbb{Z}$ such that
$$x = p^n \frac{u}{v}$$
with some integers u and v indivisible by p. Set $|x|_p = n$. Show that
$$d_p(x, y) := \begin{cases} 0 & \text{when } x = y, \\ \left(\dfrac{1}{p}\right)^{|x-y|_p} & \text{otherwise} \end{cases}$$
is an ultrametric distance on $\mathbb{Q}$, also known as the *p-adic* metric.

Problem 9.12 Let (X, d) be an ultrametric space (see Problem 9.11). Show that

(i) every triangle in X is an isosceles triangle;

(ii) every point inside a ball is its center.

Problem 9.13 Let (X, d) be a metric space. Show that if $\{x_n\}$ and $\{y_n\}$ are Cauchy sequences of X, then $\{d(x_n, y_n)\}$ is a Cauchy sequence in $\mathbb{R}$, which implies that $\{d(x_n, y_n)\}$ is convergent.

Problem 9.14 Show that every convergent sequence in a metric space (X, d) is a Cauchy sequence.

Problem 9.15 Give an example of a metric space (X, d) and a Cauchy sequence $\{x_n\} \subseteq X$ such that $\{x_n\}$ does not converge in X.

Problem 9.16 Let (X, d) be a metric space. Show that X consists of one point iff any bounded sequence in X is convergent.

Problem 9.17 Let (X, d) be a metric space. Let $\{x_n\}$ be a sequence in X such that any subsequence of $\{x_n\}$ has a subsequence which converges to some fixed point $x \in X$. Show that $\{x_n\}$ converges to x.

CHAPTER 9. METRIC SPACES

Problem 9.18 Let $\mathcal{P}$ be the set of all polynomials (of all degrees) defined on $[0,1]$. Define $d(x,y)$ by
$$d(x,y) = \max_{0 \le t \le 1} |x(t) - y(t)|.$$
Prove that $(\mathcal{P}, d)$ is not a complete metric space.

Problem 9.19 The open ball $B(x;r)$ in a metric space (X,d) is defined by
$$B(x;r) := \{y \in X : d(x,y) < r\}.$$
$B(x;r)$ is called the *unit ball* if $r = 1$. Draw the unit balls centered at $(0,0)$ in $\mathbb{R}^2$ with respect to the metrics

(a) $d_1(x,y) = \sqrt{(x_1 - y_1)^2 + (x_2 - y_2)^2}$

(b) $d_2(x,y) = |x_1 - y_1| + |x_2 - y_2|$

(c) $d_3(x,y) = \max(|x_1 - y_1|, |x_2 - y_2|)$

Problem 9.20 Show that $B(x;r)$ in a Euclidean space is convex.

Problem 9.21 In a metric space (X,d), given a ball $B(x_0;r)$, show that for any $x \in B(x_0;r)$, $B(x;s) \subseteq B(x_0;r)$ for all $0 < s \le r - d(x, x_0)$.

Problem 9.22 Describe a closed ball, open ball, and sphere with center x_0 and radius r in a metric space with the discrete metric.

Problem 9.23 Given a metric space (X,d) and a nonempty bounded subset A, the real number $\delta(A) := \sup\{d(x,y) : x, y \in A\}$ is called the *diameter* of A. It is clear that $\delta(S(x_0;r)) \le 2r$. Show that equality is not always valid.

Problem 9.24 Let $\{I_n\}$ be a sequence of bounded nonempty closed subsets of a complete metric space (X,d) such that

(a) $I_{n+1} \subseteq I_n$, for all $n \ge 1$;

(b) $\lim_{n \to \infty} \delta(I_n) = 0$, where $\delta(A) = \sup\{d(x,y) : x, y \in A\}$.

Show that $\bigcap_{n \ge 1} I_n$ is not empty and reduced to a single point.

Solutions

Solution 9.1

(a) Since $|x-y| = |y-x|$, (M2) is satisfied. Note that
$$|x-y| = 0 \Leftrightarrow x-y = 0 \Leftrightarrow x = y,$$
(M1) is satisfied. To show (M3), we will make use of the inequality $|a+b| \leq |a| + |b|$ for any $a, b \in \mathbb{R}$. Indeed, we have
$$d(x,y) = |x-y| = |(x-z) + (z-y)| \leq |x-z| + |z-y| = d(x,z) + d(z,y)$$
for any $x, y, z \in \mathbb{R}$.

(b) Since $\sqrt{|x-y|} = \sqrt{|y-x|}$, (M2) is satisfied. As in part (a),
$$\sqrt{|x-y|} = 0 \Leftrightarrow |x-y| = 0 \Leftrightarrow x = y,$$
so (M1) is satisfied. Note that if $a, b \geq 0$, then $a \geq b$ if and only if $a^2 \geq b^2$. Thus, showing (M3) is equivalent to showing $(\sqrt{|x-y|})^2 \leq (\sqrt{|x-z|} + \sqrt{|z-y|})^2$. So
$$\begin{aligned}
(\sqrt{|x-y|})^2 &= |x-y| \\
&\leq |x-z| + |z-y| \\
&\leq |x-z| + |z-y| + 2\sqrt{|x-z|}\sqrt{|z-y|} \\
&= \left(\sqrt{|x-z|} + \sqrt{|z-y|}\right)^2
\end{aligned}$$
which completes the proof of (M3).

Solution 9.2

(M1) and (M2) follow straight from the definition. For (M3), suppose $x, y, z \in X$.

- If $x = y$, then $d(x,y) \leq d(x,z) + d(z,y)$ is clear.
- If $x \neq y$, then either $z \neq x$ or $z \neq y$:
  - If $z \neq x$, then $d(x,y) = 1 = d(x,z) \leq d(x,z) + d(z,y)$.
  - If $z \neq y$, then $d(x,y) = 1 = d(y,z) \leq d(x,z) + d(z,y)$.

Solution 9.3

We show that (M1), (M2), and (M3) hold:

(M1) $d(\mathbf{x}, \mathbf{y}) = 0 \Leftrightarrow \sqrt{(x_1-y_1)^2 + (x_2-y_2)^2} = 0 \Leftrightarrow x_1 = y_1$ and $x_2 = y_2 \Leftrightarrow \mathbf{x} = \mathbf{y}$.

(M2) $d(\mathbf{x}, \mathbf{y}) = \sqrt{(x_1-y_1)^2 + (x_2-y_2)^2} = \sqrt{(y_1-x_1)^2 + (y_2-x_2)^2} = d(\mathbf{y}, \mathbf{x})$.

CHAPTER 9. METRIC SPACES

(M3) Let $\mathbf{x} = (x_1, x_2)$, $\mathbf{y} = (y_1, y_2)$, and $\mathbf{z} = (z_1, z_2)$ be in $\mathbb{R}^2$. Note that the quadratic function $\alpha t^2 + \beta t + \gamma$ has a constant sign if and only if $\beta^2 - 4\alpha\gamma \leq 0$ which is equivalent to $|\beta| \leq 2(\alpha\gamma)^{1/2}$. Now observe that the following function (of t) is never negative:

$$F(t) = [(x_1 - z_1)t + (z_1 - y_1)]^2 + [(x_2 - z_2)t + (z_2 - y_2)]^2.$$

But $F(t) = \alpha t^2 + \beta t + \gamma$ where

$$\begin{cases} \alpha &= (x_1 - z_1)^2 + (x_2 - z_2)^2, \\ \beta &= 2\Big[(x_1 - z_1)(z_1 - y_1) + (x_2 - z_2)(z_2 - y_2)\Big], \\ \gamma &= (z_1 - y_1)^2 + (z_2 - y_2)^2. \end{cases}$$

So, we must have $|\beta| \leq 2(\alpha\gamma)^{1/2}$. But

$$(\alpha\gamma)^{1/2} = 2\left\{[(x_1 - z_1)^2 + (x_2 - z_2)^2][(z_1 - y_1)^2 + (z_2 - y_2)^2]\right\}^{1/2} = 2d(\mathbf{x}, \mathbf{z})d(\mathbf{z}, \mathbf{y}),$$

so

$$\Big|(x_1 - z_1)(z_1 - y_1) + (x_2 - z_2)(z_2 - y_2)\Big| \leq d(\mathbf{x}, \mathbf{z})d(\mathbf{z}, \mathbf{y}).$$

Therefore,

$$\begin{aligned} [d(\mathbf{x}, \mathbf{y})]^2 &= (x_1 - y_1)^2 + (x_2 - y_2)^2 \\ &= [(x_1 - z_1) + (z_1 - y_1)]^2 + [(x_2 - z_2) + (z_2 - y_2)]^2 \\ &\leq (x_1 - z_1)^2 + (x_2 - z_2)^2 + 2d(\mathbf{x}, \mathbf{z})d(\mathbf{z}, \mathbf{y}) + (z_1 - y_1)^2 + (z_2 - y_2)^2 \\ &= [d(\mathbf{x}, \mathbf{z})]^2 + 2d(\mathbf{x}, \mathbf{z})d(\mathbf{z}, \mathbf{y}) + [d(\mathbf{z}, \mathbf{y})]^2 \\ &= [d(\mathbf{x}, \mathbf{z}) + d(\mathbf{z}, \mathbf{y})]^2. \end{aligned}$$

Thus, $d(x, y) \leq d(x, z) + d(z, y)$, which completes the proof of (M3).

Note that the same proof will show that the function d defined on $\mathbb{R}^n$ by

$$d\Big((x_i), (y_i)\Big) = \sqrt{(x_1 - y_1)^2 + \cdots + (x_n - y_n)^2}$$

is a distance on $\mathbb{R}^n$ also known as the *Euclidean distance*.

$\boxed{\text{Solution 9.4}}$

Before we check that (M1)–(M3) hold, note that if $\mathbf{x} = (x_1, x_2)$ and $\mathbf{y} = (y_1, y_2)$, then we have the inequality $|x_1 - y_1| \leq d(\mathbf{x}, \mathbf{y})$. This inequality will be useful in the proof of (M1)–(M3).

(M1) Assume $d(\mathbf{x}, \mathbf{y}) = 0$. If $x_2 \neq y_2$, then we have

$$d(\mathbf{x}, \mathbf{y}) = |x_1| + |x_2 - y_2| + |y_1| = 0$$

which implies $|x_2 - y_2| = 0$ or $x_2 = y_2$, which is a clear violation with our assumption. So we must have $x_2 = y_2$. In this case, we have

$$d(\mathbf{x}, \mathbf{y}) = |x_1 - y_1| = 0$$

which implies $x_1 = y_1$. In other words, we have $\mathbf{x} = \mathbf{y}$.

(M2) We observe that

$$d(\mathbf{x},\mathbf{y}) = \begin{cases} |x_1 - y_1| & \text{if } x_2 = y_2 \\ |x_1| + |x_2 - y_2| + |y_1| & \text{if } x_2 \neq y_2 \end{cases}$$
$$= \begin{cases} |y_1 - x_1| & \text{if } y_2 = x_2 \\ |y_1| + |y_2 - x_2| + |x_1| & \text{if } y_2 \neq x_2 \end{cases}$$
$$= d(\mathbf{y},\mathbf{x}).$$

(M3) Let $\mathbf{x} = (x_1, x_2)$, $\mathbf{y} = (y_1, y_2)$, and $\mathbf{z} = (z_1, z_2)$. We examine two cases:

(1) If $x_2 = y_2$, then

$$d(\mathbf{x},\mathbf{y}) = |x_1 - y_1| \leq |x_1 - z_1| + |z_1 - y_1| \leq d(\mathbf{x},\mathbf{z}) + d(\mathbf{z},\mathbf{y}).$$

(2) If $x_2 \neq y_2$, then $z_2 \neq x_2$ or $z_2 \neq y_2$. Without loss of generality, assume $z_2 \neq x_2$. Then

$$d(\mathbf{x},\mathbf{y}) = |x_1| + |x_2 - y_2| + |y_1|$$
$$\leq |x_1| + |x_2 - z_2| + |z_2 - y_2| + |y_1|$$
$$\leq \begin{cases} (|x_1| + |x_2 - z_2| + |z_1|) + |z_1 - y_1| & \text{if } y_2 = z_2 \\ (|x_1| + |x_2 - z_2| + |z_1|) + (|z_1| + |z_2 - y_2| + |y_1|) & \text{if } y_2 \neq z_2 \end{cases}$$
$$= d(\mathbf{x},\mathbf{z}) + d(\mathbf{z},\mathbf{y}).$$

Solution 9.5

We check that (M1)–(M3) hold:

(M1) First note that $d(x,y) = \max\{|x(t) - y(t)| : t \in [a,b]\} = 0$ if and only if for any $t \in [a,b]$ we have $|x(t) - y(t)| = 0$, which implies $x(t) = y(t)$. So $d(x,y) = 0$ if and only if $x(t) = y(t)$ for all $t \in [a,b]$, i.e., $x = y$.

(M2) $d(x,y) = \max\{|x(t) - y(t)| : t \in [a,b]\} = \max\{|y(t) - x(t)| : t \in [a,b]\} = d(y,x)$.

(M3) Let $x, y, z \in X$. For any $s \in [a,b]$ we have

$$|x(s) - y(s)| \leq |x(s) - z(s)| + |z(s) - y(s)| \leq \max_{t \in [a,b]} |x(t) - z(t)| + \max_{t \in [a,b]} |z(t) - y(t)|.$$

Since s was arbitrarily taken in $[a,b]$, we have

$$\max_{t \in [a,b]} |x(t) - y(t)| \leq \max_{t \in [a,b]} |x(t) - z(t)| + \max_{t \in [a,b]} |z(t) - y(t)|,$$

or $d(x,y) \leq d(x,z) + d(z,y)$.

Solution 9.6

The proofs of (M1), (M2), and (M3) follow exactly the proofs given in the previous problem.

Solution 9.7

We check that (M1) through (M3) hold:

CHAPTER 9. METRIC SPACES 189

(M1) The key behind the proof of (M1) is the following property $\int_a^b |f(t)| dt = 0$ if and only if $f(t) = 0$ for all $t \in [a,b]$ provided $f(x)$ is a continuous function. Hence for any $x, y \in X$, we have
$$d(x,y) = \int_a^b |x(t) - y(t)| \, dt = 0 \quad \Leftrightarrow \quad |x(t) - y(t)| = 0 \text{ for all } t \in [a,b],$$
which easily implies $x = y$.

(M2) For any $x, y \in X$, we have
$$d(x,y) = \int_a^b |x(t) - y(t)| \, dt = \int_a^b |y(t) - x(t)| \, dt = d(y,x).$$

(M3) For $x, y, z \in X$, we have
$$\begin{aligned} d(x,y) &= \int_a^b |x(t) - y(t)| \, dt \\ &= \int_a^b \Big(|x(t) - z(t)| + |y(t) - z(t)| \Big) \, dt \\ &\leq \int_a^b |x(t) - z(t)| \, dt + \int_a^b |y(t) - z(t)| \, dt \\ &= d(x,z) + d(y,z). \end{aligned}$$

Solution 9.8

From (M3) we get the inequalities
$$d(x,z) \leq d(x,y) + d(y,z) \quad \Rightarrow \quad d(x,z) - d(y,z) \leq d(x,y)$$
and
$$d(y,z) \leq d(y,x) + d(x,z) \quad \Rightarrow \quad d(y,z) - d(x,z) \leq d(y,x).$$
Together, these inequalities imply
$$\Big| d(x,z) - d(z,y) \Big| \leq d(x,y).$$
For the other inequality note that we have
$$\Big| d(x,y) - d(y,z) \Big| \leq d(x,z)$$
and
$$\Big| d(y,z) - d(z,w) \Big| \leq d(y,w).$$
Hence
$$\Big| d(x,y) - d(z,w) \Big| \leq \Big| d(x,y) - d(y,z) \Big| + \Big| d(y,z) - d(z,w) \Big| \leq d(x,z) + d(y,w).$$

Solution 9.9

Before we check that (M1), (M2) and (M3) hold, note that ρ is well defined since $1 + d(x,y) \neq 0$.

(M1) $\rho(x,y) = 0$ iff the numerator $d(x,y) = 0$, which is true iff $x = y$.

(M2) It is clear from the symmetry of d that $\rho(x,y) = \rho(y,x)$.

(M3) Let a, b, and c be positive numbers such that $a \leq b+c$. Then we have
$$a \leq b+c \leq b+c+2bc+abc$$
which implies
$$a + a(b+c) + abc \leq a(b+c) + b + c + 2bc + 2abc.$$
But
$$a + a(b+c) + abc = a[1+b][1+c]$$
and
$$a(b+c) + b + c + 2bc + 2abc = b[1+a][1+c] + c[1+a][1+b].$$
Hence
$$a[1+b][1+c] \leq b[1+a][1+c] + c[1+a][1+b]$$
which implies
$$\frac{a}{1+a} \leq \frac{b}{1+b} + \frac{c}{1+c}.$$
Therefore, if we set $a = d(x,y)$, $b = d(x,z)$, and $c = d(z,y)$, we know that $a \leq b+c$ since d obeys the triangle inequality. Hence
$$\frac{a}{1+a} \leq \frac{b}{1+b} + \frac{c}{1+c},$$
or
$$\rho(x,y) \leq \rho(x,z) + \rho(z,y).$$

Solution 9.10

Let us check that (M1), (M2) and (M3) hold.

(M1) From the definition of d_p we know that $d_p(x,y) = 0$ iff $x = y$.

(M2) Since $d(x,p) + d(p,y) = d(y,p) + d(p,x)$, we get $d_p(x,y) = d_p(y,x)$.

(M3) Let $x, y, z \in X$. Without loss of generality, we may assume that the three points x, y, and z are different. Then
$$d_p(x,y) = d(x,p) + d(p,y) \leq d(x,p) + d(p,z) + d(z,p) + d(p,y) = d_p(x,z) + d_p(z,y).$$

Solution 9.11

(i) Take $\mathbf{x} = (1,0,\ldots,0)$, $\mathbf{y} = (0,1,0,\ldots,0)$, and $\mathbf{O} = (0,\ldots,0)$. Then we have $d(\mathbf{x},\mathbf{y}) = \sqrt{2}$, $d(\mathbf{x},\mathbf{O}) = 1$, and $d(\mathbf{y},\mathbf{O}) = 1$. Hence
$$d(\mathbf{x},\mathbf{y}) \not\leq \max\{d(\mathbf{x},\mathbf{O}), d(\mathbf{y},\mathbf{O})\},$$
which implies that d is not an ultrametric on $\mathbb{R}^n$.

CHAPTER 9. METRIC SPACES

(ii) First let us show that the p-adic function d_p is a metric on $\mathbb{Q}$. Note that (M1) and (M2) hold directly from the definition of d_p. In order to prove (M3), we will prove

$$d_p(x,y) \leq \max\left\{d_p(x,z), d_p(y,z)\right\}$$

for any $x, y, z \in \mathbb{Q}$. This will prove (M3) and the ultrametric property at the same time. Indeed, this will follow from the inequality

$$\max\left\{d_p(x,z), d_p(y,z)\right\} \leq d_p(x,z) + d_p(y,z).$$

Let $x, y, z \in \mathbb{Q}$ be three different rationals. Set $n = |x - z|_p$, and $m = |z - y|_p$. Without loss of generality, assume $n \leq m$. By definition of $|\cdot|_p$, we have $x - z = p^n \frac{u}{v}$, and $z - y = p^m \frac{u^*}{v^*}$, which implies

$$x - y = p^n\left(\frac{u}{v} + p^{m-n}\frac{u^*}{v^*}\right) = p^n\left(\frac{uv^* + p^{m-n}vu^*}{vv^*}\right).$$

By definition of $|\cdot|_p$, we get $n \leq |x-y|_p$ since vv^* is indivisible by p. Since

$$\left(\frac{1}{p}\right)^{|x-y|_p} \leq \left(\frac{1}{p}\right)^n = \max\left\{\left(\frac{1}{p}\right)^n, \left(\frac{1}{p}\right)^m\right\}$$

or

$$d_p(x,y) \leq \max\left\{d_p(x,z), d_p(y,z)\right\}.$$

Solution 9.12

(i) Let $x, y, z \in X$. Let us prove that

$$d(x,y) = d(y,z), \text{ or } d(x,z) = d(y,z), \text{ or } d(x,y) = d(z,x).$$

Assume not. Without loss of generality assume $d(x,y) < d(y,z) < d(x,z)$. But these inequalities will contradict the fact

$$d(x,z) \leq \max\{d(x,y), d(y,z)\}.$$

(ii) Let $r > 0$ and $x, y \in X$. Assume that $d(x,y) < r$. Then $B(x;r) = B(y;r)$, where $B(x;r) = \{z \in X; d(x,z) < r\}$. We will only prove that $B(y;r) \subset B(x;r)$. Let $z \in B(y;r)$. Since

$$d(x,z) \leq \max\{d(x,y), d(y,z)\} < r,$$

we get $z \in B(x;r)$.

Solution 9.13

From Problem 9.8, we know that for any $n, m \in \mathbb{N}$ we have

$$\left|d(x_n, y_n) - d(x_m, y_m)\right| \leq d(x_n, x_m) + d(y_n, y_m).$$

So, given $\varepsilon > 0$, there exist $N_1 \in \mathbb{N}$ and $N_2 \in \mathbb{N}$ such that for any $n, m \geq N_1$ we have

$$d(x_n, x_m) < \frac{\varepsilon}{2},$$

and for any $n, m \geq N_2$ we have

$$d(y_n, y_m) < \frac{\varepsilon}{2}.$$

Set $N = \max\{N_1, N_2\}$. Then for any $n, m \geq N$ we have

$$\left| d(x_n, y_n) - d(x_m, y_m) \right| \leq d(x_n, x_m) + d(y_n, y_m) < \frac{\varepsilon}{2} + \frac{\varepsilon}{2} = \varepsilon.$$

This implies that $\{d(x_n, y_n)\}$ is a Cauchy sequence in $\mathbb{R}$.

Solution 9.14

Let $\{x_n\}$ be a convergent sequence in X with limit x. Then for each $\varepsilon > 0$ there exists some N such that $d(x_n, x) < \varepsilon/2$ for all $n \geq N$. Using the triangle inequality (M3), it follows that

$$\begin{aligned} d(x_n, x_m) &\leq d(x_n, x) + d(x, x_m) \\ &< \frac{\varepsilon}{2} + \frac{\varepsilon}{2} = \varepsilon \end{aligned}$$

for all $n, m \geq N$. Hence, $\{x_n\}$ is a Cauchy sequence.

Solution 9.15

Take $X = \mathbb{Q}$ with the metric $d(x, y) = |x - y|$. Consider the sequence $\{x_n\}$ defined in Problem 3.19 by $x_0 = 1$ and

$$x_{n+1} = \frac{1}{2}\left(x_n + \frac{2}{x_n}\right)$$

for all $n \geq 1$. Then $x_n \in \mathbb{Q}$ for all $n \in \mathbb{N}$ and $\lim x_n = \sqrt{2} \in \mathbb{R} \setminus \mathbb{Q}$, as shown in Problem 3.19. Thus, by the previous problem we have that $\{x_n\}$ is a Cauchy sequence in $\mathbb{R}$, and hence in $\mathbb{Q}$ too, but $\lim x_n \notin \mathbb{Q}$.

Solution 9.16

Obviously, if X consists of one point, then any sequence in X is constant and therefore is convergent. Conversely, assume that any bounded sequence in X is convergent. Let us prove that X consists of one point. Assume not. Let $x, y \in X$ with $x \neq y$. Consider the sequence $\{x_n\}$ defined by

$$x_{2n} = x \text{ and } x_{2n+1} = y.$$

It is clear that $\{x_n\}$ is bounded and is not convergent. Contradiction.

Solution 9.17

Assume not. Then there exists $\varepsilon_0 > 0$ such that for any $N \geq 1$, there exists $n \geq N$ with $|x_n - x| \geq \varepsilon_0$. By induction, we construct a subsequence $\{x_{n_k}\}$ of $\{x_n\}$ such that for any $k \geq 1$ we have $|x_{n_k} - x| \geq \varepsilon_0$. It is clear no subsequence of $\{x_{n_k}\}$ will converge to x. Contradiction.

Solution 9.18

It is clear that $(\mathcal{P}, d)$ is a metric space (see Problem 9.5). To show $(\mathcal{P}, d)$ is not complete, consider the following sequence:

$$x_n(t) = \sum_{k=0}^{n} \left(\frac{t}{2}\right)^k = 1 + \frac{t}{2} + \cdots + \frac{t^n}{2^n} \quad 0 \le t \le 1.$$

Clearly, $x_n(t) \in \mathcal{P}$ for each $n \in \mathbb{N}$. Next we show that the sequence $\{x_n\}$ is a Cauchy sequence. Taking $m < n$ we observe

$$\begin{aligned}
d(x_n, x_m) &= \max_{0 \le t \le 1} |x_n(t) - x_m(t)| \\
&= \max_{0 \le t \le 1} \left| \sum_{k=0}^{n} \left(\frac{t}{2}\right)^k - \sum_{k=0}^{m} \left(\frac{t}{2}\right)^k \right| \\
&= \max_{0 \le t \le 1} \left| \sum_{k=m+1}^{n} \left(\frac{t}{2}\right)^k \right| \\
&\le \max_{0 \le t \le 1} \left| \sum_{k=m+1}^{n} \frac{1}{2^k} \right| \\
&= \frac{1}{2^m} - \frac{1}{2^n}.
\end{aligned}$$

This difference is arbitrarily small for large enough m and n, which implies that $\{x_n\}$ is a Cauchy sequence in $\mathcal{P}$. However, this sequence does not converge in $(\mathcal{P}, d)$, because the only candidate for $\lim x_n(t) = \dfrac{2}{2-t}$, for $0 \le t \le 1$, and this is not a polynomial function. Since not every Cauchy sequence converges in $(\mathcal{P}, d)$, we have that $(\mathcal{P}, d)$ is not complete.

Solution 9.19

See the following figure.

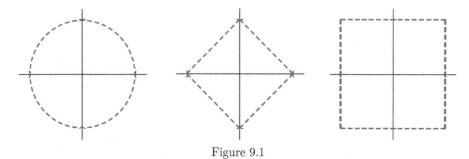

Figure 9.1

Solution 9.20

Let $y_1, y_2 \in B(x;r)$ and set $z = \alpha y_1 + (1-\alpha)y_2$, where $0 \leq \alpha \leq 1$. Then we have

$$\begin{aligned} d(x,z) &= d(x, \alpha y_1 + (1-\alpha)y_2) \\ &\leq d(x, \alpha y_1) + d(x, (1-\alpha)y_2) \\ &= \alpha d(x, y_1) + (1-\alpha)d(x, y_2). \end{aligned}$$

Since $y_1, y_2 \in B(x;r)$, we have that $d(x,y_1) < r$ and $d(x,y_2) < r$, which implies $d(x,z) < r$, and therefore $z \in B(x;r)$.

Solution 9.21

Let $y \in B(x;s)$, then $d(x,y) < s$ and by the triangle inequality we have

$$d(y, x_0) \leq d(y, x) + d(x, x_0) < s + d(x, x_0) \leq r.$$

Solution 9.22

Let $x_0 \in X$, then the sphere centered at x_0 with radius 1 is

$$S(x_0; 1) = \{x \in X : d(x_0, x) = 1\} = X \setminus \{x_0\}.$$

For $r > 0$ and $r \neq 1$, $S(x_0; r) = \{x \in X : d(x_0, x) = r\} = \emptyset$. Let $B[x_0; r]$ and $B(x_0; r)$ denote the closed and open ball with center x_0 and radius r (respectively). Then we have that

$$B[x_0; r] = B(x_0; r) = B[x_0; 1] = \begin{cases} X & \text{for } r > 1, \\ x_0 & \text{for } 0 < r < 1. \end{cases}$$

Solution 9.23

Let (X, d) be a discrete metric space where X has more than one element. Then

$$S(x_0; 1) = X \setminus \{x_0\} \quad \text{and} \quad \delta(S(x_0; 1)) = 1 < 2.$$

For $r > 0$ and $r \neq 1$, $S(x_0; r) = \emptyset$ and $\delta(S(x_0; r))$ is not defined.

Solution 9.24

Since $\{I_n\}$ is a sequence of nonempty sets, form a sequence $\{x_n\}$ with $x_n \in I_n$ for all $n \geq 1$. We claim that $\{x_n\}$ is a Cauchy sequence in X. Indeed, let $\varepsilon > 0$ and since $\lim\limits_{n \to \infty} \delta(I_n) = 0$, there exists $n_0 \geq 1$ such that for all $n \geq n_0$ we have $\delta(I_n) < \varepsilon$. Let $m \geq n \geq n_0$, then $x_n, x_m \in I_m$ because $\{I_n\}$ is decreasing. Then

$$d(x_n, x_m) \leq \delta(I_n) < \varepsilon.$$

This proves our claim. Since X is complete we conclude that $\{x_n\}$ is convergent. Let $x \in X$ be its limit. We claim that $\bigcap\limits_{n \geq 1} I_n = \{x\}$. Indeed, for any $n \geq 1$ and any $m \geq n$ we have $x_m \in I_n$.

CHAPTER 9. METRIC SPACES

Since the subsequence $\{x_m\}_{m \geq n}$ also converges to x and I_n is closed, we conclude that $x \in I_n$. Since n was arbitrary, we get $x \in \bigcap_{n \geq 1} I_n$. Hence $\bigcap_{n \geq 1} I_n$ is not empty. Let $y \in \bigcap_{n \geq 1} I_n$. By definition of the diameter, we get $d(x,y) \leq \delta(I_n)$ for all $n \geq 1$. Since $\lim_{n \to \infty} \delta(I_n) = 0$, we get $d(x,y) \leq 0$ or $d(x,y) = 0$ which implies $y = x$. This completes the proof of our statement.

Chapter 10

Fundamentals of Topology

A linguist would be shocked to learn that if a set is not closed this does not mean that it is open, or again that "E is dense in E" does not mean the same thing as "E is dense in itself."

John Edensor Littlewood (1885–1977)

- We say $A \subset \mathbb{R}$ is *open* if for every $x \in A$ there exists $\varepsilon > 0$ such that $(x - \varepsilon, x + \varepsilon) \subseteq A$. A is *closed* if its complement A^c is open. Similarly a set A in a metric space (M, d) is called *open* if for each $x \in A$, there exists an $\varepsilon > 0$ such that $B(x; \varepsilon) \subset A$. Here,

$$B(x; \varepsilon) = \{y \in M : d(x, y) < \varepsilon\}$$

is the ε-ball (also called ε-neighborhood) around x.

- Let A be a subset of a metric space (M, d) and $x \in M$. We say x is an *accumulation point* of A if every open set U containing x contains some point $y \in A$ with $y \neq x$.

- Let $A \subset (M, d)$. We say $x \in M$ is a *limit point* of a set A provided $U \cap A \neq \emptyset$ for every neighborhood U of x.

- A set A is closed in a metric space (M, d) if and only if the accumulation points of A belong to A and we set $\overline{A} := A \cup \{x \in M : x \text{ is an accumulation point of } A\}$.

- A subset A in a metric space M is called *compact* if one of the following equivalent conditions is satisfied:

a) Every open cover of A has a finite subcover.

b) Every sequence in A has a convergent subsequence converging to a point in A (sequential compactness).

Furthermore, if $A \in \mathbb{R}^n$, the above two conditions are equivalent to saying that A is closed and bounded (Heine–Borel Theorem).

- A metric space (M, d) is called *totally bounded* if for each $\varepsilon > 0$ there is a finite set $\{x_1, x_2, \cdot, x_k\}$ in M such that
$$A \subset \bigcup_{i=1}^{k} B(x_i; \epsilon).$$

- A metric space (M, d) is compact if and only if M is complete and totally bounded.

- A subset A of a metric space (M, d) is called *not connected* if there are disjoint open sets U and V such that $A \subseteq U \cup V$ and $A \cap U \neq \emptyset \neq A \cap V$. Otherwise, the set A is said to be *connected*.

- A subset A of a metric space (M, d) is said to be *path connected* if for each pair of points x and y in A, there is a path in A connecting x to y, i.e., there is a continuous function $\psi : [0, 1] \to A$ such that $\psi(0) = x$ and $\psi(1) = y$.

- Let M be a metric space and A be a subset of M, and $f : A \to \mathbb{R}$ be a continuous function. Suppose $B \subset A$ is connected and $x, y \in B$. Then for every real number c such that $f(x) < c < f(y)$, there exists a point $z \in B$ such that $f(z) = c$.
Notice that, since intervals (open or closed) are connected, the above statement is a generalized version of the *Intermediate Value Theorem* given for intervals.

Problem 10.1 Show that $\mathbb{R}$ and $\emptyset$ are both closed and open.

Problem 10.2 Let $A = \{x \in \mathbb{R} : x \text{ is irrational}\}$. Is A closed?

Problem 10.3 Prove that arbitrary unions and finite intersections of open sets are open. Using De Morgan's laws (Problem 1.9), state a corresponding result for closed sets. Give examples to prove that infinite intersections of open sets need not be open and infinite unions of closed sets need not be closed.

Problem 10.4 Show that a subset $A \subseteq \mathbb{R}$ is open if and only if A is the union of a countable collection of open intervals.

CHAPTER 10. FUNDAMENTALS OF TOPOLOGY

Problem 10.5 Prove that

1. If A is open and B is closed, then $A\backslash B$ is open and $B\backslash A$ is closed.

2. Let A be open and let B be an arbitrary subset of $\mathbb{R}$. Is AB necessarily open?
   (Here $AB = \{xy \in \mathbb{R} : x \in A \text{ and } y \in B\}$.)

Problem 10.6 Prove that an open ball in any metric space (X, d) is open.

Problem 10.7 Prove that a subset A of any metric space (X, d) is open if and only if A is the union of open balls.

Problem 10.8 Prove that any closed subset A of any metric space (X, d) is a countable intersection of open sets.

Problem 10.9 Let A be a subset of (M, d). Prove that

1. $\overline{A} = \bigcap \{C; A \subset C \text{ and } C \text{ is closed}\}$;

2. $A = \overline{A}$ if and only if A is closed.

Problem 10.10 Show that if $x \in (M, d)$ is an accumulation point of A, then x is a limit point of A. Is the converse true?

Problem 10.11 Prove that an element $x \in (M, d)$ is a limit point of A if and only if there is a sequence $\{x_n\} \subset A\backslash\{x\}$ such that $\lim_{n \to \infty} x_n = x$.

Problem 10.12 For a subset $A \subset (M, d)$, prove the following are equivalent:

1. A is closed.

2. A contains all its limit points.

3. Every sequence in A which converges in (M, d) has its limit in A.

Problem 10.13 Consider $\mathbb{R}^n$ endowed with the Euclidean distance. Show that if $B(a; r) = \{x \in \mathbb{R}^n : d(x, a) < r\}$, then

$$\overline{B}(a; r) = \{x \in \mathbb{R}^n : d(x, a) \leq r\}.$$

Is this conclusion true in any metric space?

Problem 10.14 Let A be a subset of a metric space (M, d). Define

$$\check{A} := \{a \in A : a \text{ is an interior point of } A\} \text{ and } \text{int}(A) := \bigcup\{O \subseteq A : O \text{ is open in } M\}.$$

Show that

1. $\check{A} = \text{int}(A)$.

2. A is open if and only if $A = \text{int}(A)$.

Problem 10.15 For a given set A in a metric space (M, d), the *boundary* of A, ∂A, is defined to be the set $\partial A = \overline{A} \cap \overline{(M \setminus A)} = \overline{A} \cap \overline{(A^c)}$. Prove or answer the following:

1. ∂A is a closed set.

2. $\partial A = \partial(M \setminus A)$.

3. If $x \in \partial A$, does x have to be a limit point?

4. $x \in \partial A$ if and only if for every $\varepsilon > 0$, $B(x; \varepsilon)$ contains points of A and of $M \setminus A$.

Problem 10.16 Discuss whether the following sets are open or closed. Determine the interior, closure, and boundary of each set.

1. $(1, 2)$ in $\mathbb{R}$.

2. $[1, 2]$ in $\mathbb{R}$.

3. $\bigcap_{n=1}^{\infty} \left[-2, \dfrac{1}{n}\right)$ in $\mathbb{R}$.

4. $(0, 1) \cap \mathbb{Q}$ in $\mathbb{R}$.

Problem 10.17 Show that for a set $A \subset (M, d)$, $x \in \overline{A}$ if and only if there is a sequence $(x_k) \in A$ with $x_k \to x$.

Problem 10.18 For a subset A of a metric space, show that $x \in \overline{A}$ if and only if

$$d(x, A) = \inf\{d(x, y) : y \in A\} = 0.$$

Problem 10.19 Prove the following:

1. If x is a accumulation point of a set $A \subseteq \mathbb{R}$, then every open set containing x contains infinitely many points of A. In particular, A must be infinite.

2. Let $A \subset \mathbb{R}$ be nonempty and bounded above. Let $x = \sup(A)$. Show that $x \in \overline{A}$ and $x \in \partial A$.

Problem 10.20 Which of the following sets are compact?

1. $[0,1] \cup [5,6] \subset \mathbb{R}$
2. $\{x \in \mathbb{R} : x \geq 0\} \subset \mathbb{R}$
3. $\{x \in \mathbb{R} : 0 \leq x \leq 1 \text{ and } x \text{ is irrational}\}$
4. $A = \{1, \frac{1}{2}, \frac{1}{3}, \ldots, \frac{1}{n}, \ldots\} \cup \{0\}$

Problem 10.21 Prove that a closed subset of a compact set is compact.

Problem 10.22 Let K be a nonempty compact set. Let $\{A_n\}$ be a nonempty decreasing sequence of closed subsets of K. Prove that $\bigcap_{n \geq 1} A_n$ is not empty. (Cantor's Intersection Theorem)

Problem 10.23 Let (M,d) be a metric space. Let K be a nonempty subset of M which is sequentially compact, i.e., any sequence $\{x_n\}$ in K has a subsequence which converges to a point in K. Show that for any arbitrary open cover $\{O_\alpha\}$ of K, there exists $\varepsilon > 0$ such that for any $x \in K$, there exists α such that $B(x, \varepsilon) \subset O_\alpha$.

Problem 10.24 Let (M,d) be a metric space. Show that $K \subset M$ is compact iff K is sequentially compact, i.e., any sequence $\{x_n\}$ in K has a subsequence which converges to a point in K.

Problem 10.25 Give an example of a closed and bounded set in a metric space which is not compact.

Problem 10.26 Show that a totally bounded set is bounded. Give an example of a bounded set which is not totally bounded.

Problem 10.27 Let (M, d_M) and (N, d_N) be two metric spaces. Let $A \subset M$ and $B \subset N$ be two compact subsets. Prove that $A \times B$ is compact in $(M \times N, d)$, where $d = d_M + d_N$.

Problem 10.28 Let $S_1 = [0, \frac{1}{3}] \cup [\frac{2}{3}, 1]$ be obtained from $[0,1]$ by removing the middle third $(\frac{1}{3}, \frac{2}{3})$. Repeat the process to obtain $S_2 = [0, \frac{1}{9}] \cup [\frac{2}{9}, \frac{1}{3}] \cup [\frac{2}{3}, \frac{7}{9}] \cup [\frac{8}{9}, 1]$. In general, S_{n+1} is obtained from S_n by removing the middle third of each interval in S_n. Let $C = \bigcap_{n \geq 1} S_n$, also known as the Cantor set. Prove that

1. C is compact.

2. $\text{int}(C) = \emptyset$.

3. C has infinitely many points.

4. The total length of the intervals removed is equal to 1.

Problem 10.29 Prove that for a function f mapping $A \subseteq \mathbb{R}^n$ into $\mathbb{R}^m$, the following are equivalent:

1. f is continuous on A.

2. $f^{-1}(U) = \{x \in \mathbb{R}^n : f(x) \in U\}$ is open in A for every open set U in $\mathbb{R}^m$.

This property is sometimes called the *topological characterization of continuity*. Also, the above equivalence is also true if we replace $\mathbb{R}^n$ and $\mathbb{R}^m$ with arbitrary metric spaces.

Problem 10.30 Give an example of a continuous function $f : \mathbb{R} \to \mathbb{R}$ and an open set U in $\mathbb{R}$ such that $f(U)$ is not an open set.

Problem 10.31 Prove that if $f : M \to N$ is continuous and A is a compact subset of M, then $f(A)$ is a compact subset of N (i.e., the continuous image of a compact set is compact).

Problem 10.32 Prove that a continuous real-valued function defined on a compact set is bounded and it assumes maximum and minimum values.

Problem 10.33 Let (M, d) be a metric space with a nonempty compact subset A. Prove that for every $x_0 \in M$, there exists a $y_0 \in A$ such that

$$d(x_0, y_0) = d(x_0, A) = \inf\{d(x_0, y) : y \in A\}.$$

Problem 10.34 Prove that a continuous one-to-one mapping T from a compact metric space (M, d) onto a metric space (N, d') is a homeomorphism.

Note that a map T between two metric spaces $T : (M, d) \to (N, d')$ is called a *homeomorphism* if T is one-to-one continuous and has a continuous inverse.

Problem 10.35 Let A be a compact set and let $f: A \to A$ be a continuous function with the property that for all $x, y \in A$, $d(f(x), f(y)) \geq d(x, y)$. ($f$ is expanding)

1. Prove that f is one-to-one and the inverse map $f^{-1}: f(A) \to A$ is continuous.

2. Show that $f(A) = A$.

Problem 10.36 Let X be a closed subset of $\mathbb{R}$ and let $K(X)$ denote the collection of all nonempty compact subsets of X. Show that

$$d_H(A, B) := \max\left\{\sup_{a \in A} d(a, B), \sup_{b \in B} d(b, A)\right\}$$

defines a metric on $K(X)$, where $d(x, A) = \inf_{a \in A} d(x, a)$.

Problem 10.37 (Generalized Cantor's Intersection Theorem)
The set measure of noncompactness $\alpha(D)$ for a bounded subset D of (M, d) is defined as

$$\alpha(D) = \inf\left\{r > 0: \ D \subset \bigcup_{i=1}^{n} A_i \ \ \operatorname{diam}(A_i) \leq r\right\}.$$

Show that

a) If D is compact, then $\alpha(D) = 0$.

b) $D_1 \subset D_2 \implies \alpha(D_1) \leq \alpha(D_2)$ (α is monotone).

c) $\alpha(\overline{D}) = \alpha(D)$ (invariant when given the closure).

d) If $\{F_n\}$ is a decreasing sequence of nonempty closed and bounded subsets of a complete metric space (M, d) and if $\lim_{n \to \infty} \alpha(F_n) = 0$, then the intersection of all the F_n is nonempty and compact.

Problem 10.38 A subset A of a metric space (M, d) is called *not connected* if there are disjoint open sets U and V such that $A \subseteq U \cup V$, $A \cap U \neq \emptyset$, and $A \cap V \neq \emptyset$. Otherwise, the set A is said to be *connected*. Show that the following sets are not connected:

1. $\mathbb{Q} \subset \mathbb{R}$, the set of rational numbers

2. $\mathbb{N}$, the set of natural numbers

Problem 10.39 Let C be the Cantor set as defined in Problem 10.28. Prove that C is totally disconnected, that is, if $x, y \in C$ and $x \neq y$, then $x \in U$ and $y \in V$, where U and V are open sets that disconnect C.

Problem 10.40 Prove that the continuous image of a connected set is connected.

Problem 10.41 Prove that the continuous image of a path-connected set is path connected.

Problem 10.42 A subset A of a metric space (M, d) is said to be *path connected* if for each pair of points x and y in A, there is a path in A connecting x to y, i.e., there is a continuous function $f : [0, 1] \to A$ such that $f(0) = x$ and $f(1) = y$. Prove that every path-connected set is connected.

Problem 10.43 Show that the set $A = \{x \in \mathbb{R}^n : \| x \| \leq 1\}$ is compact and connected.

Problem 10.44 Give an example of a set which is connected but not path connected.

Problem 10.45 Prove that $A \subseteq \mathbb{R}$ is connected if and only if it is an interval (bounded or unbounded). Then show that if $f : M \to \mathbb{R}$ is continuous, then $f(M)$ is an interval, where M is a connected metric space. In particular, f takes on every value between any two given function values. (Generalized Intermediate Value Theorem)

Problem 10.46 Let $A \subseteq M$ and $B \subseteq N$ be path connected. Prove that $A \times B$ is path connected in $M \times N$.

Problem 10.47 (Baire's Theorem) Let $\{O_n\}_{n \in \mathbb{N}}$ be a sequence of dense open subsets of $\mathbb{R}$. Show that $\bigcap_{n \in \mathbb{N}} O_n$ is dense in $\mathbb{R}$.

Problem 10.48 (Baire's Category Theorem) Prove that if M is a nonempty, complete metric space, then it is of second category itself.
Notice that if $M = \mathbb{R}$, since $\mathbb{R}$ is complete, the above question is equivalent to proving that the set of real numbers $\mathbb{R}$ cannot be written as a countable union of nowhere dense sets.

Problem 10.49 Use Baire's Category Theorem to show that the set of all rationals $\mathbb{Q}$ is not the intersection of a countable collection of open sets. Use this result to show that the set of irrationals is not the union of a countable collection of closet sets.

Problem 10.50 Let C be a subset of $\mathbb{R}$. Define the *Cantor–Bendixson derivative* of C, denoted C', by
$$C' = \{x \in \mathbb{R}; \ x \text{ is a limit point or accumulation point of } C\}.$$
Show C' is closed, and if $C' \neq \emptyset$, then C is infinite.

Problem 10.51 A subset P of $\mathbb{R}$ is said to be perfect if and only if $P' = P$, where P' is the Cantor–Bendixson derivative of P. Show that any nonempty perfect subset of $\mathbb{R}$ is not countable.

Problem 10.52 A point a is a condensation point of $A \subset \mathbb{R}$ if for any $\varepsilon > 0$, the set $(a - \varepsilon, a + \varepsilon) \cap A$ is infinite not countable. Set $P = \{x \in \mathbb{R};\ x \text{ is a condensation point of } A\}$. Show that P is either empty or perfect.

Problem 10.53 Let C be a closed subset of $\mathbb{R}$. Show that $C = P \cup F$, where P is perfect, F is countable, and $P \cap F = \emptyset$. This is known as the *Cantor–Bendixson Theorem*.

Solutions

Solution 10.1

The set $\mathbb{R}$ itself is open because for any $x \in \mathbb{R}$, $(x-1, x+1) \subseteq \mathbb{R}$. The empty set $\emptyset$ is also open "vacuously." $\mathbb{R}$ and $\emptyset$ must also both be closed since $\mathbb{R}^c = \emptyset$ and $\emptyset^c = \mathbb{R}$.

Solution 10.2

No. Note that $0 \in A^c$. For any $\varepsilon > 0$, the set $(0-\varepsilon, 0+\varepsilon)$ contains irrational points, such as $\dfrac{\sqrt{2}}{n}$ for large integers n. Thus A^c is not open, and therefore A is not closed. Similarly one can show that A is not open as well.

Solution 10.3

Let ϑ be a collection of open sets. We want to prove that $\bigcup_{G \in \vartheta} G$ is open. Set $\bigcup_{G \in \vartheta} G = A$ and let $x \in A$, then $x \in G$ for some $G \in \vartheta$. Since G is open, there is an $\varepsilon > 0$ such that $(x-\varepsilon, x+\varepsilon) \subset G$. But $G \subseteq A$, so $(x-\varepsilon, x+\varepsilon)$ is also contained in A. Thus, A is open.

Let $\{G_1, G_2, \ldots, G_n\}$ be a finite collection of open sets, and let $B = \bigcap_{1 \le k \le n} G_k$. If $B = \emptyset$, then it is open by the previous problem. Suppose $B \ne \emptyset$. Let $x \in B$, then $x \in G_k$ for each k. Each G_k is open, so there are ε-neighborhoods $(x-\varepsilon_k, x+\varepsilon_k)$ satisfying $(x-\varepsilon_k, x+\varepsilon_k) \subseteq G_k$ for each k. Let $\varepsilon = \min\{\varepsilon_1, \varepsilon_2, \ldots, \varepsilon_n\}$, then $(x-\varepsilon, x+\varepsilon) \subseteq (x-\varepsilon_k, x+\varepsilon_k) \subset G_k$ for all k. Hence $(x-\varepsilon, x+\varepsilon) \subseteq B$. Thus, B is open.

By De Morgan's laws, we have

$$\left(\bigcap_{i \in I} G_i\right)^c = \bigcup_{i \in I} G_i^c \quad \text{and} \quad \left(\bigcup_{i=1}^n G_i\right)^c = \bigcap_{i=1}^n G_i^c.$$

Therefore we can say finite unions and arbitrary intersections of closed sets are closed.

Since

$$\bigcap_{n=1}^\infty \left(-\frac{1}{n}, \frac{1}{n}\right) = \{0\},$$

we conclude that infinite intersections of open sets need not be open. By taking the complement we get

$$\bigcup_{n=1}^\infty \left(-\frac{1}{n}, \frac{1}{n}\right)^c = \mathbb{R}\backslash\{0\},$$

which shows that infinite unions of closed sets need not be closed.

Solution 10.4

Since the union of any collection of open sets is open, we only need to show that if A is open, then A is a countable union of open intervals. Then for all $x \in A$, there exists $\varepsilon > 0$ such that $(x-\varepsilon, x+\varepsilon) \subseteq A$. Now find rationals $r_x \in (x-\varepsilon, x)$ and $s_x \in (x, x+\varepsilon)$. Then clearly $x \in (r_x, s_x)$ and $A = \bigcup_{x \in A} (r_x, s_x)$. Notice that the number of open intervals with rational end points is less than or equal to $\mathbb{Q}X\mathbb{Q}$, where $\mathbb{Q}$ is the set of rational numbers. Since $\mathbb{Q} \times \mathbb{Q}$ is countable, we see that this is a countable union.

Solution 10.5

1. $A \setminus B = A \cap B^c$ and the intersection of a finite number of open sets is open. Similarly, $B \setminus A = B \cap A^c$ and the intersection of closed sets is closed.

2. If A is an open subset of $\mathbb{R}$ and $B = \{0\}$, then $AB = \{0\}$, which is not open. So, in general, the answer to the question is "No." However, under certain conditions, the answer is "Yes." For example, if A is open and $y \neq 0$, then $yA = \{y\}A$ is an open subset of $\mathbb{R}$, and since $AB = \bigcup_{y \in B} yA$, AB is open because arbitrary unions of open sets are open.

Solution 10.6

Let $B(x,r) = \{y \in X; d(x,y) < r\}$, for $x \in X$ and $r > 0$. Let us prove that any point $y \in B(x,r)$ is an interior point, i.e., there exists $\varepsilon > 0$ such that $B(y, \varepsilon) \subset B(x, r)$. Set $\varepsilon = r - d(x, y)$. Since $y \in B(x, r)$, we have $\varepsilon > 0$. Let $z \in B(y, \varepsilon)$. Then

$$d(x,z) \leq d(x,y) + d(y,z) < d(x,y) + \varepsilon = r$$

which proves $z \in B(x,r)$. Hence $B(y, \varepsilon) \subset B(x, r)$. This completes the proof of our claim.

Solution 10.7

Note first that the union of open balls is open since an open ball is also an open set (see the previous problem). In order to complete the proof of our statement, we are only left to prove that any open set A of X is the union of open balls. By definition of open sets, for any $a \in A$, there exists $\varepsilon_a > 0$ such that $B(a, \varepsilon_a) \subset A$. Obviously we have

$$A \subseteq \bigcup_{a \in A} \{a\} \subseteq \bigcup_{a \in A} B(a, \varepsilon_a) \subseteq A$$

which yields $\bigcup_{a \in A} B(a, \varepsilon_a) = A$.

Solution 10.8

For each $n \geq 1$, set $O_n = \bigcup_{a \in A} B(a, 1/n)$. Obviously O_n is open and $A \subset O_n$ for all $n \geq 1$. Therefore, $A \subset \bigcap_{n \geq 1} O_n$. We claim that in fact we have $A = \bigcap_{n \geq 1} O_n$. Assume not. Then there exists $y \in \bigcap_{n \geq 1} O_n \setminus A$. Since A is closed, then A^c is open and contains y. Hence there exists $\varepsilon > 0$ such that $B(y, \varepsilon) \subset A^c$. Since $\varepsilon > 0$, there exists $n \geq 1$ such that $1/n < \varepsilon$. Because $y \in O_n$, there exists $a \in A$ such that $d(a, y) < 1/n < \varepsilon$. So $B(y, \varepsilon) \cap A$ is not empty, which implies $A \cap A^c \neq \emptyset$. Contradiction.

Solution 10.9

1. Let C be a closed subset of M such that $A \subset C$. Let us show that $\overline{A} \subset C$. Let $x \in \overline{A}$. If $x \in A$, then we have $x \in C$. Assume x is a limit point of A. Assume x is not in C. Since C is closed, there must exist $\varepsilon > 0$ such that $B(x, \varepsilon) \cap C = \emptyset$. On the other hand, we have $B(x, \varepsilon) \cap A$ contains a point different from x since x is a limit of A. That point is also in $B(x, \varepsilon) \cap C$ since $A \subset C$. So $C \cap C^c \neq \emptyset$. Contradiction. Hence $\overline{A} \subset C$ which implies

$$\overline{A} \subset \bigcap \{C; A \subset C \text{ and } C \text{ is closed}\}.$$

In order to finish the proof, let us show that $\overline{A}$ is closed. Let $x \in \overline{A}^c$, then $x \in A^c$ and x is not a limit point of A. Hence there exists $\varepsilon > 0$ such that $B(x, \varepsilon) \cap A = \emptyset$. In particular we have $B(x, \varepsilon) \subset A^c$. Any $y \in B(x, \varepsilon)$ will not be in A and is not a limit point of A. Indeed since $B(x, \varepsilon)$ is open, there must exist $\delta > 0$ such that $B(y, \delta) \subset B(x, \varepsilon) \subset A^c$ which implies $B(y, \delta) \cap A = \emptyset$. Therefore $y \in \overline{A}^c$ or $B(x, \varepsilon) \subset \overline{A}^c$. This shows that $\overline{A}^c$ is open or $\overline{A}$ is closed. Then we have

$$\overline{A} \subset \bigcap \{C; A \subset C \text{ and } C \text{ is closed}\} \subset \overline{A},$$

which completes the proof of our statement.

2. If A is closed, then obviously we have $A = \overline{A}$ from the above result. Conversely, if $A = \overline{A}$, then A is closed since $\overline{A}$ is closed.

Solution 10.10

An accumulation point must be a limit point since

$$U \cap (A \setminus \{x\}) \subseteq U \setminus A.$$

If x is an accumulation point of A and U is an open set containing x, then $U \cap (A \setminus \{x\}) \neq \emptyset$ and therefore $U \setminus A \neq \emptyset$. The converse is not true, a limit point need not be an accumulation point. For example, if we set $A = [0, 1) \cup \{3\} \subseteq \mathbb{R}$, $3 \in A$ so it is a limit point, but if we take $U = (1.5, 4)$ an open set containing 3 contains no other points $y \in A$ with $y \neq 3$.

Notice that 3 is an isolated point of A, a limit point either belongs to A or is an accumulation point of A, it could be an isolated point of A; an accumulation point is not.

Solution 10.11

Let x be a limit point of A. For each $n \geq 1$, choose some element $x_n \neq x$ in $B(x, \frac{1}{n})$. Then $\{x_n\}$ is a sequence in $A \setminus \{x\}$ and since $1/n \to 0$, then $x_n \to x$. Conversely, assume there exists $\{x_n\}$ is a sequence in $A \setminus \{x\}$ which converges to x. Then for each $\varepsilon > 0$, there exists $n \in \mathbb{N}$ such that $x_n \in B(x, \varepsilon)$, i.e., $x_n \in (A \setminus \{x\}) \cap B(x, \varepsilon)$. Therefore each neighborhood of x contains an element of A other than x. Thus x is a limit point of A.

Solution 10.12

- To show that (a)⇒(b), notice that $A = \overline{A}$ since A is closed (see Problem 10.9). Hence any limit point of A is in A.

- To show that (b)⇒(c), let $\{x_n\}$ be a sequence in A which converges to x. Let us prove that $x \in A$. Assume not, then $x \notin A$. Let us show that x is a limit point of A. Indeed let $\varepsilon > 0$. Then there exists $n \in \mathbb{N}$ such that $x_n \in B(x, \varepsilon)$. Since $x \notin A$ and $x_n \in A$, $B(x, \varepsilon) \cap A$ contains a point different from x. This proves x is a limit point of A. So $x \in A$ from our assumption. Contradiction.

- To show (c)⇒(a), assume (c) holds and (a) does not. Hence A^c is not open. Then there exists $x \in A^c$ such that for any $\varepsilon > 0$, we have $B(x, \varepsilon) \cap A \neq \emptyset$. In particular, for any $n \geq 1$, there exists $x_n \in B(x, 1/n) \cap A$. The sequence $\{x_n\}$ converges to x and $\{x_n\}$ is in A. Our assumption implies $x \in A$. Contradiction.

Solution 10.13

If $x \in \overline{B}(a;r) \setminus B(a,r)$, then x is a limit point of $B(a,r)$. Hence there is a sequence $\{x_k\} \subset B(a;r)$ such that $x_k \to x$. We want to show $d(x,a) \leq r$. Suppose not. So we have $d(x,a) > r$. Then there exists ε such that $d(x,a) - \varepsilon > r$. Since $x_k \to x$, there exists $k \in \mathbb{N}$ such that $d(x, x_k) < \varepsilon$. By the triangle inequality we get

$$d(a,x) \leq d(a, x_k) + d(x, x_k) < d(a, x_k) + \varepsilon,$$

which implies $r \leq d(a,x) - \varepsilon < d(a, x_k)$ which contradicts the fact $x_k \in B(a,r)$. Thus if $x \in \overline{B}(a;r)$, then $d(a,x) \leq r$, i.e., $\overline{B}(a;r) \subset \{x \in \mathbb{R} : d(x,a) \leq r\}$. Note that we could have obtained this inclusion easily by showing that $\{x \in \mathbb{R} : d(x,a) \leq r\}$ is closed. Now suppose $d(a,x) \leq r$, so now we want to show that $x \in \overline{B}(a;r)$. Let $x_n = a + (1 - \frac{1}{n})(x - a)$, then

$$d(x_n, a) \leq \left(1 - \frac{1}{n}\right) d(x,a) \leq \left(1 - \frac{1}{n}\right) r < r.$$

So $x_n \in B(a;r)$ for all n. Since $x_n \to x$, then $x \in \overline{B}(a;r)$, i.e., $\{x \in \mathbb{R} : d(x,a) \leq r\} \subset \overline{B}(a;r)$. This completes the proof of the first part.

This characterization of the closure of open balls may not be true in any metric space. Indeed, take $\mathbb{R}$ endowed with the discrete distance d (see Problem 9.23). Then

$$B(0,1) = \{x \in \mathbb{R}; d(0,x) < 1\} = \{0\}$$

is open and closed, while $\{x \in \mathbb{R}; d(0,x) \leq 1\} = \mathbb{R}$.

Solution 10.14

1. Let $a \in \overset{\smile}{A}$. Then there is an $\varepsilon > 0$ such that $B(a;\varepsilon) \subseteq A$. Since $B(a;\varepsilon)$ is open, $B(a;\varepsilon) \subseteq \text{int}(A)$, which proves $a \in \text{int}(A)$, i.e., $\overset{\smile}{A} \subseteq \text{int}(A)$. Now let $a \in \text{int}(A)$. Then there is an open subset $O \subseteq A$ such that $a \in O$. Since O is open, there exists $\varepsilon > 0$ such that $B(a,\varepsilon) \subset O$. Hence $B(a,\varepsilon) \subset A$ which means a is an interior point of A, so $\text{int}(A) \subseteq \overset{\smile}{A}$.

2. Note that int(A) is open because it is a union of open sets. Hence $A = \text{int}(A)$ which will force A to be open. Assume A is open, then $A = \mathring{A}$ from the definition of open sets. Using the above property, we get $A = \text{int}(A)$.

Solution 10.15

1. Since the intersection of two closed sets is closed, it follows that ∂A is a closed set.

2. This follows from the definition of the boundary and the fact that $M\setminus(M\setminus A) = A$.

3. No. Indeed let $A = \{0\} \subseteq \mathbb{R}$. Then A has no limit points, but $\partial A = \{0\}$.

4. Let $x \in \partial A$. Since $\partial A = \overline{A} \cap \overline{(M\setminus A)}$, we have that $x \in \overline{A}$ and $x \in \overline{(M\setminus A)}$. Note that if $x \in \overline{K}$, for any set K, then for any $\varepsilon > 0$, we have $B(x,\varepsilon) \cap K \neq \emptyset$. Indeed, by definition of $\overline{K}$, we know that either $x \in K$ which obviously will imply $B(x,\varepsilon) \cap K \neq \emptyset$, or x is a limit point of K. And in this case using the definition of limit points, we get again $B(x,\varepsilon) \cap K \neq \emptyset$. So since $x \in \overline{A} \cap \overline{(M\setminus A)}$, we get $B(x,\varepsilon) \cap A \neq \emptyset$ and $B(x,\varepsilon) \cap (M\setminus A) \neq \emptyset$.
Conversely, let $x \in M$ such that for any $\varepsilon > 0$, $B(x;\varepsilon)$ contains points of A and of $M\setminus A$. Let us prove that $x \in \partial A$. It is enough to prove that $x \in \overline{A}$. If $x \in A$, then we have nothing to show. Assume $x \notin A$. Let $\varepsilon > 0$. We know that $B(x,\varepsilon) \cap A \neq \emptyset$, since $x \notin A$, then $B(x,\varepsilon) \cap A$ contains a point different from x. Hence x is a limit point of A, i.e., $x \in \overline{A}$.

Solution 10.16

1. $(1,2)$ is an open subset of $\mathbb{R}$ therefore $A = \text{int}(A) = (1,2)$. The endpoints 1 and 2 are limit points, so they are in the closure: $\overline{A} = [1,2]$. Since $\overline{\mathbb{R}\setminus A} = \{x : x \leq 1\} \cup \{x : x \geq 2\}$, the boundary of A, $\partial A = \overline{A} \cap \overline{\mathbb{R}\setminus A} = \{1,2\}$.

2. $[1,2] = \{x \in \mathbb{R} : 1 \leq x \leq 2\}$ is a closed subset of $\mathbb{R}$. Indeed the complement of $[1,2]$ in $\mathbb{R}$ is the union of the two open half-lines, namely: $\mathbb{R}\setminus[1,2] = \{x : x < 1\} \cup \{x : x > 2\}$. As the union of two open subsets is open, $\mathbb{R}\setminus[1,2]$ is open, so $[1,2]$ is closed. Since $[1,2]$ is closed, then $\overline{[1,2]} = [1,2]$, $\text{int}(B) = (1,2)$, and $\partial B = \{1,2\}$.

3. $\bigcap_{n=1}^{\infty}[-2,\frac{1}{n}) = [-2,0]$. This is because if $-2 \leq x \leq 0$, then $x \in [-2,\frac{1}{n})$ for any positive n, so x is in the intersection. On the other hand, for any $x > 0$, we can find an integer $n > 0$ (Archimedean Principle) with $0 < \frac{1}{n} < x$. Therefore $x \notin [-2,\frac{1}{n})$ for that n, and so $x \notin \bigcap_{n=1}^{\infty}[-2,\frac{1}{n})$.
The interval $[-2,0]$ is closed by the same argument used in part 2. If we set $C = [-2,0]$, then, as above, $\text{int}(C) = (-2,0)$, $\overline{C} = [-2,0]$, and $\partial C = \{-2,0\}$.

4. If we set $F = (0,1) \cap \mathbb{Q}$, F is not open in $\mathbb{R}$, since, for example, every open interval centered at $\frac{1}{2}$ contains irrational numbers. It is also not closed, since the complement is not open, as any open interval around 0 contains rational numbers $\frac{1}{n}$ for large enough positive n in the integers. Thus, F is neither closed nor open.

If $r \in F$ and $\varepsilon > 0$, then the interval $(r-\varepsilon, r+\varepsilon)$ contains irrational numbers, and so r cannot be an interior point, and thus $\text{int}(F) = \emptyset$. Now if $y \in [0,1]$, then for any $\varepsilon > 0$, there are rational numbers in $(y-\varepsilon, y+\varepsilon)$, so y is a limit point of F. Thus we have $\overline{F} = [0,1]$. By similar reasoning, one can show $\overline{\mathbb{R}\setminus F} = \mathbb{R}$. Therefore, $\partial F = \overline{F} \cap \overline{\mathbb{R}\setminus F}) = [0,1] \cap \mathbb{R} = [0,1]$.

Solution 10.17

$\Rightarrow$: Assume $x \in \overline{A}$. then either $x \in A$ or $x \in \{$accumulation points of $A\}$. If $x \in A$, form a sequence $\{x, x, \dots\} \to x$. If $x \in \{$accumulation points of $A\}$, then for every $\varepsilon > 0$ there exist $y \neq x$ such that $y \in B(x; \epsilon) \cap A$. Setting $\epsilon_n = \frac{1}{n}$ and choosing $x_k \neq x$ from each $B(x; \epsilon_n) \cap A$ we obtain the desired sequence. Note that $(x_k) \to x$, because for all $\varepsilon > 0$, we can choose $N = \dfrac{1}{\varepsilon}$, so that $d(x_k, x) < \varepsilon$ whenever $k \geq N$.

$\Leftarrow$: Assume there is a sequence $(x_k) \in A$ converging to x. If $x \in A$, then $x \in \overline{A}$, since $A \subset \overline{A}$. Suppose x is not in A, we must show $x \in \{$accumulation points of $A\}$. Since $(x_k) \to x$, for any $\varepsilon > 0$ there exists N such that $x_k \in B(x; \varepsilon)$ whenever $k \geq N$. Since we know $(x_k) \in A$, we have $x_k \in B(x; \varepsilon) \cap A$, this implies that x is an accumulation point.

Solution 10.18

Let $x \in \overline{A}$. If $x \in A$, then obviously we have $d(x, A) = 0$. Assume $x \notin A$. Then x is a limit point of A. Thus, for any $\varepsilon > 0$, there exists a $y \in B(x, \varepsilon) \cap A$, i.e., $d(x, y) < \varepsilon$. Therefore $d(x, A) < \varepsilon$, for any $\varepsilon > 0$. Hence $d(x, A) = 0$. Conversely suppose $d(x, A) = 0$. If $x \in A$, then $x \in \overline{A}$. Assume $x \notin A$. Then by the property of the infimum, for any $\varepsilon > 0$, there is a $y \in A$ such that $d(x, y) < \varepsilon$, i.e., $y \in B(x, \varepsilon) \cap A$. Since $x \notin A$, then $y \neq x$. Therefore, x is a limit point of A, and thus $x \in \overline{A}$.

Solution 10.19

1. Proof is done by contradiction. Suppose there is an open set O containing x and containing only a finite number of points of A different from x. Say $x_1, x_2, \dots, x_n$ are the points of A in O other than x. Let $\varepsilon_1 = \min\{d(x, x_1), d(x, x_2), \dots, d(x, x_n)\}$. Also, since O is open, we know there exists some $\varepsilon_2 > 0$ such that $B(x; \varepsilon_2) \subseteq O$. Set $\varepsilon = \min\{\varepsilon_1, \varepsilon_2\}$. Notice that $\varepsilon > 0$ and $B(x; \varepsilon)$ contains no points of A other than x. This contradicts the fact that x is a limit point of A.

2. If $\sup A \in A$, then $\sup A \in \overline{A}$. Assume that $\sup A \notin A$. Then by the properties of supremum, for every $\varepsilon > 0$, there is a $y \in A$ such that $\sup A - \varepsilon < y \leq \sup A < \sup A + \varepsilon$, i.e., $y \in A \cap (\sup A - \varepsilon, \sup A + \varepsilon)$. Thus, $\sup A$ is a limit point of A, so $\sup A \in \overline{A}$.

   Let $\varepsilon > 0$. By the properties of supremum, there is an element $y \in A$ such that $\sup A - \varepsilon < y \leq \sup A$, i.e., every interval $(\sup A - \varepsilon, \sup A + \varepsilon)$ contains a point of A. On the other hand, the upper half of such an interval $(\sup A, \sup A + \varepsilon) \subset \mathbb{R} \backslash A$, since $\sup A$ is an upper bound of A. Thus $\sup A \in \overline{\mathbb{R} \backslash A}$. Therefore, $\sup A \in \overline{A} \cap \overline{(\mathbb{R} \backslash A)} = \partial A$, as desired.

Solution 10.20

1. Compact, because it is closed and bounded (Heine–Borel Theorem).

2. Noncompact, because it is unbounded.

3. Noncompact, because if we set $A = \{x \in \mathbb{R} : 0 \leq x \leq 1 \text{ and } x \text{ is irrational}\}$, then $x = \frac{1}{3} \notin A$, but every interval around it contains irrational numbers which are in A. Therefore A^c is not open, so A is not closed, and thus A cannot be compact.

4. Let $\{O_\alpha\}$ be an arbitrary open cover of A. The point 0 lies in one of the open sets, suppose $0 \in O_{\alpha_0}$ for some α_0. Since O_{α_0} is open and $\frac{1}{n} \to 0$, there is an N such that $\frac{1}{N}, \frac{1}{N+1}, \ldots$ all lie in O_{α_0}. Now since $\{O_\alpha\}$ is an open cover of A, we know there must exist $O_{\alpha_1}, O_{\alpha_2}, \ldots, O_{\alpha_N}$ (not necessarily all distinct from one another, but we can rename and give multiple names to the sets as is needed) such that $1, 1/2, \ldots, 1/N \subset O_{\alpha_1} \cup O_{\alpha_2} \cup \ldots \cup O_{\alpha_N}$. Then $\{O_{\alpha_0}, O_{\alpha_1}, O_{\alpha_2}, \ldots, O_{\alpha_N}\}$ is a finite subcover of A. Therefore, every open cover has a finite subcover, so A is compact.

Solution 10.21

Let K be a compact set and $F \subset K$ be a closed subset. Let $\{O_\alpha\}$ be an arbitrary open cover of F. Then $\{F^c, O_\alpha\}$ is an open cover of K. Since K is compact, there exists a finite subcover $\{F^c, O_{\alpha_1}, O_{\alpha_2}, \ldots, O_{\alpha_N}\}$ of K. It is easy to check that $\{O_{\alpha_1}, O_{\alpha_2}, \ldots, O_{\alpha_N}\}$ is a finite subcover of F. Thus F is compact.

Solution 10.22

Assume not. Then $\bigcap_{n \geq 1} A_n = \emptyset$, which easily implies $K \subset \bigcup_{n \geq 1} A_n^c$. Since A_n is closed, then $\{A_n^c\}$ is an open cover of K. Because K is compact, there exist $n_1, \ldots, n_N$ such that $K \subset A_{n_1}^c \cup \ldots \cup A_{n_N}^c$ which implies $A_{n_1} \cap \ldots \cap A_{n_N} = \emptyset$. Since $\{A_n\}$ is decreasing we have $A_{\max\{n_i\}} \subset A_{n_1} \cap \ldots \cap A_{n_N}$. This will contradict the fact

$$\emptyset \neq A_{\max\{n_i\}} \subset A_{n_1} \cap \ldots \cap A_{n_N} = \emptyset.$$

Solution 10.23

Assume not. Then for any $\varepsilon > 0$, there exists $x_\varepsilon \in K$ such that $B(x_\varepsilon, \varepsilon) \not\subset O_\alpha$, for any α. In particular, there exists a sequence $\{x_n\}$ in K such that $B(x_n, 1/n) \not\subset O_\alpha$, for any $n \geq 1$ and any α. Our assumption on K ensures the existence of a subsequence $\{x_{n_k}\}$ of $\{x_n\}$ which converges to some $x \in K$. Since $\{O_\alpha\}$ covers K, there exists α_0 such that $x \in O_{\alpha_0}$. Since O_{α_0} is open, there must exist $\varepsilon_0 > 0$ such that $B(x, \varepsilon_0) \subset O_{\alpha_0}$. But $\{x_{n_k}\}$ converges to x, then there exists n_i such that $B(x_{n_i}, 1/n_i) \subset B(x, \varepsilon_0) \subset O_{\alpha_0}$ contradicting the way the subsequence was constructed.

Solution 10.24

Assume that $K \subset M$ is nonempty and compact. Let $\{x_n\}$ be a sequence in K. Set $X_n = \overline{\{x_i, i \geq n\}}$. Then $\{X_n\}$ is a nonempty sequence of decreasing closed subsets of K. Problem 10.22 will ensure that $\bigcap_{n \geq 1} X_n \neq \emptyset$. Let x be in this intersection. Then $x \in K$ as well. Let us show that there exists a subsequence of $\{x_n\}$ which converges to x. Let $\varepsilon > 0$, then for any $n \geq 1$, $B(x, \varepsilon) \cap \{x_i, i \geq n\} \neq \emptyset$ since $x \in \overline{\{x_i, i \geq n\}}$. Set $\varepsilon = 1$, then there exists $n_1 \geq 1$ such that $x_{n_1} \in B(x, 1)$. Let $\varepsilon = 1/2$, then there exists $n_2 \geq n_1 + 1$ such that $x_{n_2} \in B(x, 1/2)$. By induction we construct a subsequence $\{x_{n_i}\}$ of $\{x_n\}$ such that $x_{n_i} \in B(x, 1/i)$, for any $i \geq 1$. Clearly $\{x_{n_i}\}$ converges to x.

Conversely assume that K is sequentially compact. Let us show that K is compact. Let us prove that for any $\varepsilon > 0$ there exists a finite set of points $a_1, \ldots, a_N$ in K such that

$$K \subset B(a_1, \varepsilon) \cup \ldots \cup B(a_N, \varepsilon).$$

CHAPTER 10. FUNDAMENTALS OF TOPOLOGY

Indeed, assume not. Then there exists $\varepsilon > 0$ such that for any finite set of points $a_1, \ldots, a_N$ in K, we have $K \not\subset B(a_1, \varepsilon) \cup \ldots \cup B(a_N, \varepsilon)$. In particular, we fix $x_1 \in K$, then there exists $x_2 \in K$ such that $d(x_1, x_2) \geq \varepsilon$. Assume $x_1, \ldots, x_n$ are known. Since $K \not\subset B(x_1, \varepsilon) \cup \ldots \cup B(x_n, \varepsilon)$, there exists $x_{n+1} \in K$ such that $d(x_{n+1}, x_i) \geq \varepsilon$ for $1 \leq i \leq n$. By induction we construct a sequence $\{x_n\}$ in K such that $d(x_i, x_j) \geq \varepsilon$ for $i \neq j$. Such sequence is called ε-separated. Obviously such sequence will not have a convergent subsequence. Contradiction. Let us complete the proof of our claim, i.e., K is compact. Indeed let $\{O_\alpha\}$ be an arbitrary open cover of K. Problem 10.23 ensures the existence of $\varepsilon > 0$ such that for any $x \in K$, there exists α such that $B(x, \varepsilon) \subset O_\alpha$. For that same ε, there exists a finite set of points $\{a_i, i = 1, \ldots, N\}$ in K such that $K \subset B(a_1, \varepsilon) \cup \ldots \cup B(a_N, \varepsilon)$. For any $i = 1, \ldots, N$, there exists α_i such that $B(a_i, \varepsilon) \subset O_{\alpha_i}$. Clearly $\{O_{\alpha_i}, i = 1, \ldots, N\}$ is a finite subcover of K.

Solution 10.25

Let M be the set of all bounded sequences in $\mathbb{R}$. Consider the distance $d(\{x_n\}, \{y_n\}) = \sup_{n \in \mathbb{N}} |x_n - y_n|$.
Consider
$$C = \left\{ \{x_n\}, \sup_{n \in \mathbb{N}} |x_n| \leq 1 \right\}.$$
Then C is not empty, closed, and bounded in M. Let us show that C is not compact. Indeed, for any $k \in \mathbb{N}$, let $e_k = \{x_n\}$, where $x_n = 0$ for any $n \neq k$ and $x_k = 1$. Then $e_k \in C$, for any $k \in \mathbb{N}$. But $d(e_i, e_j) = 1$ whenever $i \neq j$. Hence $\{e_i\}$ is 1-separated in C. So C cannot be sequentially compact (see Problem 10.24). In particular, C is not compact.

Another easier example is the discrete distance on any infinite set.

Solution 10.26

Suppose $A \subset (M, d)$ is totally bounded, then for each $\varepsilon > 0$ there is a finite set (called ε-net) $\{x_1, \ldots, x_k\}$ in M such that $A \subset \bigcup_{i=1}^{k} B(x_i, \varepsilon)$. Now observe that $B(x_i, \varepsilon) \subset B(x_1, \varepsilon + d(x_i, x_1))$, therefore if we set $R = \varepsilon + \max\{d(x_2, x_1), \ldots, d(x_k, x_1)\}$, then $A \subset B(x_1, R)$ and thus a totally bounded set is bounded. To show that a bounded set is not necessarily bounded, consider $(c_0, \| \cdot \|_\infty)$, where by c_0 we mean the space of all sequences converging to 0 and for $x = (x_n) \in c_0$ its norm is defined by $\| x \|_\infty = \sup |x_n|$. Consider the unit sphere $S(0; 1)$ of c_0, define a sequence (x_n) as $x_n = \{0, \ldots, 0, 1, 0, \ldots\}$ (1 in the nth place), then for all $n \neq m$
$$\| x_n - x_m \|_\infty = \sup |x_i - x_j| = 1.$$
Therefore for $\varepsilon \leq \frac{1}{2}$, the ball with radius ε contains a particular element of the sequence (x_n) and contains no other element of the sequence. Thus $S(0; 1)$ does not have an ε-net for $\varepsilon \leq \frac{1}{2}$ and therefore is not totally bounded.

Solution 10.27

Let us use the sequential characterization of compact sets in metric spaces (see Problem 10.24). Let a sequence $(a_n, b_n) \subseteq A \times B$. Since A is compact, we know there exists some subsequence a_{n_k} that converges to some point $a \in A$ as $k \to \infty$. Similarly, since B is compact, the subsequence (b_{n_k}) has a sub-subsequence $(b_{n_{k_l}})$ that converges to some $b \in B$ as $l \to \infty$. Thus $(a_{n_{k_l}}, b_{n_{k_l}}) \to (a, b)$ as $l \to \infty$. Since this is true for any arbitrary sequence (a_n, b_n), this proves that $A \times B$ is compact.

Solution 10.28

1. Notice that $C \subset [0,1]$, so it is bounded. $C = \bigcap_{n \geq 1} S_n$, where each S_n is a union of finitely many closed intervals and so is closed. Thus, the set C is an intersection of closed sets and so C is closed as well. Therefore, C is a closed and bounded subset of $\mathbb{R}$, so C is compact.

2. The length of each of the subintervals making up the set S_n is $\frac{1}{3^n}$, so the intersection can contain no interval longer than this. Since $\frac{1}{3^n} \to 0$ as $n \to \infty$, the intersection can contain no interval of positive length. If $a \in C$ is an interior point, we should be able to find an interval around a with positive length s. There are no such intervals, so $\text{int}(C) = \emptyset$.

3. Begin by noting that C contains the endpoints of all the intervals for each S_n and that each S_n has a total of 2^{n+1} endpoints. Since the number of endpoints in S_n goes to ∞ as $n \to \infty$, we have that C has infinitely many points.

4. By summing up the length of the deleted intervals, we are able to obtain:
$$\frac{1}{3} + \frac{2}{9} + \frac{4}{27} + \cdots = \frac{1}{3}\left[1 + \frac{2}{3} + \left(\frac{2}{3}\right)^2 + \cdots\right] = \frac{1}{3} \cdot \frac{1}{1 - \frac{2}{3}} = 1.$$

Solution 10.29

Suppose f is continuous and U is an open subset of $\mathbb{R}^m$. Let $a \in f^{-1}(U)$, then $f(a) \in U$ and since U is open, there is an $\varepsilon > 0$ such that $B(f(a); \varepsilon) \subseteq U$. From the continuity of f, there is a real number $\delta > 0$ such that
$$|f(x) - f(a)| < \varepsilon \text{ for all } x \in A, \text{ such that } |x - a| < \delta.$$
This means that $f(B(a; \delta) \cap A) \subset B(f(a); \varepsilon) \subseteq U$. Hence, $B(a; \delta) \cap A \subset f^{-1}(U)$ and therefore $f^{-1}(U)$ is open in A. This proves that (a) implies (b).

Conversely, suppose (b) holds. Fix an $a \in A$ and $\varepsilon > 0$ and take $U = B(f(a); \varepsilon)$. Then by assumption, $f^{-1}(U)$ is open in A and contains the point a, i.e., there is a real number $\delta > 0$ such that $B(a; \delta) \cap A \subset f^{-1}(U)$. In other words, $|f(x) - f(a)| < \varepsilon$ for all $x \in A$ with $|x - a| < \delta$, and therefore f is continuous.

Solution 10.30

Suppose $U = (0, 1) \subset \mathbb{R}$ and $f : \mathbb{R} \to \mathbb{R}$ is a function defined as $f(x) = 5$, then $f(U) = \{5\}$ is not an open set. Notice that we can take U as any open subset of $\mathbb{R}$ and $f(x) = k$ for any constant k.

Solution 10.31

Suppose (b_n) is a sequence in $f(A)$. For each n, choose $a_n \in A$ such that $f(a_n) = b_n$. Since A is compact, there is a subsequence (a_{n_k}) that converges to some point $a \in A$. By the continuity of f, it follows that $b_{n_k} = f(a_{n_k}) \to f(a) \in f(A)$. Therefore, any sequence (b_n) in $f(A)$ has a convergent subsequence, converging to a point in $f(A)$. Thus $f(A)$ is compact.

Solution 10.32

Let $f : M \to \mathbb{R}$ be continuous and let A be a compact subset of M. By Problem 10.31, we know that $f(A)$ is a compact subset of $\mathbb{R}$. By the Heine–Borel Theorem, $f(A)$ is closed and bounded. Thus $\sup(f(A))$ and $\inf(f(A))$ exist and belong to $f(A)$. Therefore, there exist $p, P \in A$ such that for all $a \in A$, $\inf(f(A)) = f(p) \le f(a) \le f(P) = \sup(f(A))$.

Solution 10.33

Suppose T^{-1} is not continuous at $y_0 \in N$. Then there exists $\varepsilon > 0$ and a sequence (y_n) in N such that $(y_n) \to y_0$ but the sequence (x_n) where $x_n = T^{-1} y_n$ and $x_0 = T^{-1} y_0$ has the property that

$$d(x_n, x_0) > \varepsilon \text{ for all } n \in \mathbb{N}.$$

However, since (M, d) is compact (x_n) has a convergent subsequence (x_{n_k}) which is convergent to some $x_1 \in M$. But T is continuous at x_1, so y_{n_k} where $y_{n_k} = T x_{n_k}$ converges to $y_1 = T x_1$. However, we are given that $(y_{n_k}) \to y_0$, so $y_1 = y_0$. Since T is one-to-one we have $x_1 = x_0$ but this contradicts the assumption that $d(x_n, x_0) > \varepsilon$ for all $n \in \mathbb{N}$.

Solution 10.34

Consider the real function f on A defined by $f(x) = d(x, x_0)$. Now $|f(x) - f(y)| = |d(x, x_0) - d(y, x_0)| \le d(x, y)$, so f is continuous on A. But A is compact, so f has a minimum on A (Problem 10.32). That is, there exists a $y_0 \in A$ such that

$$f(y_0) = d(x_0, y_0) = \inf\{d(x_0, y) : y \in A\} = d(x_0, A).$$

Solution 10.35

1. f is one-to-one because if $f(x) = f(y)$, then $0 \ge d(x, y)$, and therefore $x = y$. The continuity of the inverse function follows from $d(x, y) = d\left(f(f^{-1}(x)), f(f^{-1}(y))\right) \ge d(f^{-1}(x), f^{-1}(y))$.

2. Suppose there exists $x \in A$ such that $x \notin f(A)$. Since $f(A)$ is compact, by Problem 10.18, we know $d(x, f(A)) = d > 0$. Note that we have

$$d \le d(x, f^h(x)) \le d(f^n(x), f^{n+h}(x)),$$

for any $n, h \in \mathbb{N}$, where $f^n = f \circ \cdots \circ f$ n times. In particular, the sequence $\{f^n(x)\}$ is d-separated. In Problem 10.24 we showed that such sequences do not exist in sequentially compact metric spaces, contradicting the fact that $f(A)$ is compact. So we have $f(A) = A$.

Solution 10.36

Since A is closed, $d(x, A) = 0$ if and only if $x \in A$ (see Problem 10.18). Therefore $d_H(A, B) = 0$ if and only if $A = B$. It is clearly symmetric, i.e., $d_H(A, B) = d_H(B, A)$. For the triangle inequality, since A, B, C are three compact subsets of X, for each $a \in A$, by Problem 10.33, we know the existence of a closest point $b \in B$ so that $d(a, B) = d(a, b)$. Similarly, there is a closest point $c \in C$ to b with $d(b, C) = d(b, c)$. Therefore,

$$d(a, C) \leq d(a, c) \leq d(a, b) + d(b, c) = d(a, B) + d(b, C) \leq d_H(A, B) + d_H(B, C),$$

and hence, $\sup_{a \in A} d(a, C) \leq d_H(A, B) + d_H(B, C)$. Now if we reverse the roles of A and C above we can obtain that $\sup_{c \in C} d(c, A) \leq d_H(A, B) + d_H(B, C)$. Combining these two inequalities, we get that

$$d_H(A, C) \leq d_H(A, B) + d_H(B, C).$$

Solution 10.37

1. Follows from the definition.

2. If $D_1 \subset D_2$ and $D_2 \subset \bigcup_{i=1}^{n} A_i$ with $\operatorname{diam}(A_i) \leq r$, then covering for D_2 also covers D_1 and hence $\alpha(D_1) \leq \alpha(D_2)$.

3. Since $D \subset \overline{D}$ by the above part we have $\alpha(D) \leq \alpha(\overline{D})$. Conversely, if $D \subset \bigcup_{i=1}^{n} A_i$, then $\overline{D} \subset \bigcup_{i=1}^{n} \overline{A_i}$; however, $\operatorname{diam}(A_i) = \operatorname{diam}(\overline{A_i}) \leq r$ implies $\alpha(\overline{D}) \leq \alpha(D)$.

4. $\lim_{n \to \infty} \alpha(F_n) = 0$ implies that for each n, F_n is compact, by the nested intervals property of the compact nonempty sets we have $\bigcap_{n=1}^{\infty} F_n \neq \emptyset$.

Solution 10.38

1. For $x, y \in \mathbb{Q}$ with $x < y$, choose an irrational number z with $x < z < y$. Then setting $U = (-\infty, z)$ and $V = (z, \infty)$, we see that $\mathbb{Q} \subset U \cup V$, both U and V intersect with $\mathbb{Q}$, and $U \cap V = \emptyset$. Therefore, $\mathbb{Q}$ is not connected.

2. $\mathbb{N}$ is not connected, because if we set $U = \{0\}$ and $V = \{1, 2, 3, \ldots\}$, then both are open subsets of $\mathbb{N}$ (with respect to the relative topology inherited from the topology of $\mathbb{R}$) with $U \cap V = \emptyset$ and $\mathbb{N} = U \cup V$.

Solution 10.39

We use the same notations as in Problem 10.28. Let x and y be distinct points of C. Without loss of generality assume $x < y$. Then $x, y \in S_n$ for every n. Since the subintervals making up S_n each have length $1/3^n$, the points x and y must lie in different subintervals if n is large enough so that $|x - y| > 1/3^n$. At least one of the subintervals (x_n, y_n) removed from S_{n-1} to create S_n lies between x and y, i.e., $x \leq x_n < y_n \leq y$. Now pick a point z such that $x_n < z < y_n$. Thus $z \notin S_n$, so $z \notin C$. Let $U = \{a \in \mathbb{R} : a < z\}$ and $V = \{b \in \mathbb{R} : z < b\}$, so $U \cap V = \emptyset$ and both are open

CHAPTER 10. FUNDAMENTALS OF TOPOLOGY   217

sets. Moreover, $C \subset \mathbb{R}\backslash\{z\} = U \cup V$. Thus $x \in U$ and $y \in V$ with U and V disconnecting C. C is totally disconnected as claimed.

Solution 10.40

Assume not. Assume $f : X \to Y$ is continuous and X is connected, but the range $f(X)$ is *not* connected. Then there are disjoint open sets U and V in Y such that $f(X) \subseteq U \cup V$ and $f(X) \cap U \neq \emptyset \neq f(X) \cap V$. If we let $K_1 = f^{-1}(U)$ and $K_2 = f^{-1}(V)$, since f is continuous, the inverse image of an open set is open, and thus K_1 and K_2 are open subsets of X. Now $f(X) \subseteq U \cup V$ and $U \cap V = \emptyset$ implies that for each $x \in X$, $f(x) \in U$ or $f(x) \in V$ but not both. Therefore K_1 and K_2 cover X and $K_1 \cap K_2 = \emptyset$. Moreover, since $f(X) \cap U \neq \emptyset$ and $f(X) \cap V \neq \emptyset$, it follows that $K_1 \cap X \neq \emptyset \neq$ and $K_2 \cap X \neq \emptyset$. This shows that X is not connected, so we have arrived at a contradiction. Therefore, the continuous image of a connected set is connected.

Solution 10.41

Suppose $f : X \to Y$ is continuous and $C \subset X$ is a path-connected set. Then for every $x, y \in C$ there is a continuous path $\psi : [0,1] \to C$ connecting x to y, i.e., if $(t_n) \to t$ in $[0,1]$, then $\psi(t_n) \to \psi(t)$. Since composition of two continuous functions is continuous we certainly have $f(\psi(t_n)) \to f(\psi(t))$. Continuous path $f \circ \psi$ connects every two points in $f(C)$.

Solution 10.42

Suppose A is path connected, but not connected. Then there are disjoint open sets U and V such that $A \subseteq U \cup V$, $A \cap U \neq \emptyset$, and $A \cap V \neq \emptyset$. Pick points $u \in A \cap U$ and $v \in A \cap V$. Since A is path connected, there exists a continuous function $f : [0,1] \to A$ such that $f(0) = u$ and $f(1) = v$. Set $P = f([0,1])$. It follows that $P \subseteq U \cup V$, moreover $u \in P \cap U$ and $v \in P \cap V$, so both are nonempty. This shows that P is not connected. But this is a contradiction because a continuous image of the closed interval $[0,1]$, which is connected, is connected.

Solution 10.43

A is bounded because $A \subset B(0,2)$, let $x \in A^c = \{x \in \mathbb{R}^n : \|x\| > 1\}$, then $B(x, \|x\| - 1) \subset A^c$, thus A^c is open. A is both closed and bounded subset of $\mathbb{R}^n$ and thus A is compact by the Heine–Borel Theorem. To show A is connected we show A is path connected. Let $x, y \in A$, the straight line connecting x to y is the required path $\psi : [0,1] \to \mathbb{R}^n$, since

$$\|\psi(t)\| = \|(1-t)x + ty\| \leq (1-t)\|x\| + t\|y\| = 1.$$

Solution 10.44

Let $F = A \cup B = \{(0,t) : |t| \leq 1\} \cup \{(x, \sin\frac{1}{x}) : 0 < x \leq 1\}$. We claim that F is connected. Suppose U and V are two disjoint open sets such that $F \subseteq U \cup V$. The point $(0,0)$ must belong to one of them, so, without loss of generality, assume $(0,0) \in U$. Therefore U intersects the line A. Since A is path connected, $A \subseteq U$. Since U is an open set, there exists some $\varepsilon > 0$ such that the ball centered at $(0,0)$ with radius ε is contained in U. But $(0,0)$ is a limit point of B, so U contains a point of B. Since B is a graph, it is path connected, so by the same argument, $B \subseteq U$, which shows that $F \cap V = \emptyset$. Therefore, F is connected.

We prove that F is not path connected via contradiction. Suppose F is path connected. Then there exists a continuous function $f : [0,1] \to F$ such that $f(0) = (0,0)$ and $f(1) = (1, \sin 1)$. Now set $a = \sup\{x : f(x) \in A\}$. Since f is continuous, there exists a $\delta > 0$ such that $|f(x) - f(a)| < \frac{1}{2}$ if $|x - a| \leq \delta$. Set $b = a + \delta$. Let $f(a) = (0, y)$ and $f(b) = (u, \sin \frac{1}{u})$. A similar argument shows that there is a point $a < c < b$ such that $f(c) = (t, \sin \frac{1}{t})$, where $t \leq \frac{u}{1+2\pi u}$. Therefore, $f([c,b])$ is a connected subset of F containing both $f(c)$ and $f(b)$. On the other hand, the graph $G = \{(x, \sin \frac{1}{x}) : t \leq x \leq u\}$ is connected: if we remove any part of it we will disconnect it, so $G \subseteq f([c,b])$. Note that $\sin \frac{1}{x}$ takes both values ± 1 on $[t, u]$, so there is a point $w = (w, \sin \frac{1}{w})$ on G with $|f(w) - f(a)| > |\sin \frac{1}{w} - y| \geq 1$. But this is a contradiction to the above where $|f(x) - f(a)| < \frac{1}{2}$ for all $x \in [a, b]$. Therefore F is *not* path connected.

$\boxed{\text{Solution 10.45}}$

Intervals are path connected, and hence they are connected. For the converse, assume that A is not an interval and we will show that it is not connected. A is not an interval implies that there exist points x, y, z such that $x < y < z$ where $x, z \in A$ but $y \notin A$. Then by setting $U = (-\infty, y)$ and $V = (y, \infty)$, we get $A \subseteq U \cup V$, $U \cap V = \emptyset$, $A \cap U \neq \emptyset$, and $A \cap V \neq \emptyset$. Thus A is not connected. Next let M be a connected metric space and $f : M \to \mathbb{R}$ is continuous. Since the continuous image of a connected set is connected, $f(M)$ is connected. From the first part of this problem we conclude that $f(M)$ is an interval.

$\boxed{\text{Solution 10.46}}$

Since $A \times B = \{(a, b) : a \in A, b \in B\}$, if we take $x = (x_1, x_2)$ and $y = (y_1, y_2)$ in $A \times B$, then x_1 and y_1 are in the path-connected set A and x_2 and y_2 are in the path-connected set B. Hence there exist $\gamma : [0,1] \to A$ such that $\gamma(0) = x_1$ and $\gamma(1) = y_1$, and $\varphi : [0,1] \to B$ such that $\varphi(0) = x_2$ and $\varphi(1) = y_2$. Now set $f(t) = (\gamma(t), \varphi(t))$. Then $f : [0, 1] \to A \times B$ is continuous since each of the coordinate functions is continuous. Moreover we have $f(0) = x$ and $f(1) = y$. This shows that $A \times B$ is path connected.

$\boxed{\text{Solution 10.47}}$

Let $x \in \mathbb{R}$ and $\varepsilon > 0$. It is enough to show that $\bigcap_{n \in \mathbb{N}} O_n \cap (x - \varepsilon, x + \varepsilon)$ is not empty. Since O_0 is dense in $\mathbb{R}$, then $O_0 \cap (x - \varepsilon, x + \varepsilon) \neq \emptyset$ and is open. Let $x_0 \in O_0 \cap (x - \varepsilon, x + \varepsilon)$ and $\varepsilon_0 < \varepsilon/4$ such that $(x_0 - \varepsilon_0, x_0 + \varepsilon_0) \subset O_0 \cap (x - \varepsilon, x + \varepsilon)$. Then by induction one can build the sequences $\{x_n\}$ and $\{\varepsilon_n\}$ such that

1. $\varepsilon_n < \frac{\varepsilon_{n-1}}{2}$,

2. $(x_n - \varepsilon_n, x_n + \varepsilon_n) \subset O_n \cap (x_{n-1} - \varepsilon_{n-1}, x_{n-1} + \varepsilon_{n-1})$,

for any $n \geq 1$. It is easy to check that $\varepsilon_n < \varepsilon/2^{n+2}$, for any $n \in \mathbb{N}$. Condition 2 implies

$$(x_m - \varepsilon_m, x_m + \varepsilon_m) \subset (x_n - \varepsilon_n, x_n + \varepsilon_n), \; m > n.$$

In particular, we have $|x_m - x_n| < \varepsilon_n$, for any $m > n$. So $\{x_n\}$ is Cauchy. Hence $\{x_n\}$ converges to some $z \in \mathbb{R}$. Let us prove that $z \in \bigcap_{n \in \mathbb{N}} O_n \cap (x - \varepsilon, x + \varepsilon)$. Indeed, since $|x_m - x_n| < \varepsilon_n$, for any $m > n$, we get $|z - x_n| \leq \varepsilon_n$, for any $n \in \mathbb{N}$. In particular, we have

$$z \in (x_{n+1} - 2\varepsilon_{n+1}, x_{n+1} + 2\varepsilon_{n+1}) \subset (x_n - \varepsilon_n, x_n + \varepsilon_n) \subset O_n,$$

for any $n \in \mathbb{N}$. This implies $z \in \bigcap_{n \in \mathbb{N}} O_n$. On the other hand, we have $(x_n - \varepsilon_n, x_n + \varepsilon_n) \subset (x - \varepsilon, x + \varepsilon)$ which implies $z \in (x - \varepsilon, x + \varepsilon)$. Therefore we have proved that $z \in \bigcap_{n \in \mathbb{N}} O_n \cap (x - \varepsilon, x + \varepsilon)$.

The proof given here is analytical in nature and may be extended to complete metric spaces. In fact, Baire's Category Theorem extends to complete metric spaces as shown in the following problem.

Solution 10.48

Recall that subset A of a metric space M is said to be *nowhere dense* if its closure $\overline{A}$ has no interior points. (If $M = \mathbb{R}$, $\overline{A}$ contains no nonempty intervals, for example $\mathbb{Z}$ is nowhere dense in $\mathbb{R}$.) We say A is of *first category* in M if A is the union of countably many sets each of which is nowhere dense in M. A is *second category* in M if A is not first category.

Proof is done by contradiction. Suppose $M \neq \emptyset$ and first category in itself, then

$$M = \bigcup_{k=1}^{\infty} M_k$$

with each M_k nowhere dense in M. We will construct a Cauchy sequence $\{x_k\}$ whose limit x which exists by completeness is in no M_k, thus contradicting the representation above. By assumption M_1 is nowhere dense in M, which means $\overline{M_1}$ does not contain a nonempty open set. But M is open in itself, and this implies that $\overline{M_1} \neq M$. Therefore the complement of $\overline{M_1}$ in M is not empty and open. Using the definition of open set this means we can choose a point x_1 in $\overline{M_1}^c$ such that the open ball centered at x_1 and radius ε_1, $B(x_1; \varepsilon_1)$ is contained in $\overline{M_1}^c$, i.e.,

$$B(x_1; \varepsilon_1) \subset \overline{M_1}^c.$$

By assumption M_2 is nowhere dense in M, i.e., $\overline{M_2}$ does not contain a nonempty open set. Hence it does not contain the open ball $B(x_1; \frac{1}{2}\varepsilon_1)$. This implies that $\overline{M_2}^c \cap B(x_1; \frac{1}{2}\varepsilon_1) \neq \emptyset$ so we can choose a point x_2 in $\overline{M_2}^c \cap B(x_1; \frac{1}{2}\varepsilon_1)$ and $\varepsilon_2 < \frac{1}{2}\varepsilon_1$ such that

$$B(x_2, \varepsilon_2) \subset \overline{M_2}^c \cap B(x_1; \frac{1}{2}\varepsilon_1).$$

Continuing in this manner, we obtain a sequence of open balls $b(x_k; \varepsilon_k)$ such that $B(x_k; \varepsilon_k) \cap M_k \neq \emptyset$ and

$$B(x_k, \frac{1}{2}\varepsilon_k) \subset B(x_k; \varepsilon_k)$$

for $k = 1, 2, \ldots$. Furthermore $\varepsilon_k < \frac{1}{2^k}$ guarantees that the sequence formed by centers x_k of these balls forms a Cauchy sequence. Since M was complete the sequence $\{x_k\}$ converges to some $x \in M$. Also for every m with $n > m$ we have

$$B(x_n; \varepsilon_n) \subset B(x_m; \frac{1}{2}\varepsilon_m).$$

Now using the triangle inequality we get

$$d(x_m, x) \leq d(x_m, x_n) + d(x_n, x) < \frac{1}{3}\varepsilon_m + d(x_n, x)$$

which implies that $d(x_m, x) \to \frac{1}{2}\varepsilon_m$ as $n \to \infty$ thus proving that $x \in B(x_m; \varepsilon_m)$ for every m. But $B(x_m; \varepsilon_m) \subset \overline{M_m}^c$, and we showed that $x \notin M_m$ for every m, so that

$$x \notin M = \bigcup_{k=1}^{\infty} M_k.$$

This contradicts the fact that $x \in M$.

Solution 10.49

Assume not. Then there exists $\{O_n\}$ a sequence of open sets such that $\mathbb{Q} = \underset{n \in \mathbb{N}}{\cap} O_n$. Since $\mathbb{Q}$ is countable, then we may write $\mathbb{Q} = \{r_n; n \in \mathbb{N}\}$. Set $\tilde{O}_n = O_n \setminus \{r_n\}$, for $n \in \mathbb{N}$. It is clear that $\tilde{O}_n$ is open as an intersection of two open sets. Since $\mathbb{Q} \subset O_n$, then O_n is dense in $\mathbb{R}$ and consequently $\tilde{O}_n$ is also dense in $\mathbb{R}$, for any $n \in \mathbb{N}$. Baire's Category Theorem implies that $\underset{n \in \mathbb{N}}{\cap} \tilde{O}_n$ is not empty and is dense in $\mathbb{R}$. But this contradicts

$$\underset{n \in \mathbb{N}}{\cap} \tilde{O}_n = \underset{n \in \mathbb{N}}{\cap} O_n \setminus \mathbb{Q} = \emptyset.$$

Finally assume that the set of irrationals is the union of a countable collection of closed sets. Then by taking the complement one can easily prove that $\mathbb{Q}$ is the intersection of a countable collection of open sets. Contradiction.

In topology, this conclusion means that $\mathbb{Q}$ is not G_δ-set and $\mathbb{R} \setminus \mathbb{Q}$ is not an F_σ.

Solution 10.50

If $C' = \emptyset$, we have nothing to prove. So assume $C' \neq \emptyset$. Let us prove that C' is closed. Let $a \notin C'$, then a is not a limit point of C. Hence there exists $\varepsilon > 0$ such that $(a - \varepsilon, a + \varepsilon) \cap C$ does not contain a point different from a. Since $a \notin C$, we have $(a - \varepsilon, a + \varepsilon) \cap C = \emptyset$. In fact, we have $(a - \varepsilon, a + \varepsilon) \cap C' = \emptyset$. Indeed assume not, i.e., $(a - \varepsilon, a + \varepsilon) \cap C' \neq \emptyset$. Let $a^* \in (a - \varepsilon, a + \varepsilon) \cap C'$. Since $(a - \varepsilon, a + \varepsilon)$ is open, there exists $\delta > 0$ such that $(a^* - \delta, a^* + \delta) \subset (a - \varepsilon, a + \varepsilon)$. Since a^* is a limit point of C, $(a^* - \delta, a^* + \delta) \cap C \neq \emptyset$ which in turns implies $(a - \varepsilon, a + \varepsilon) \cap C \neq \emptyset$. Contradiction. Hence $\mathbb{R} \setminus C'$ is open or equivalently C' is closed. Finally let us prove that C is infinite. Assume not. So $C = \{c_1, \ldots, c_n\}$. Since C' is not empty, let $c^* \in C'$. Set $\varepsilon = \min\{|c^* - c_i|; c_i \neq c^*, i = 1, \ldots, n\}$. It is clear that $\varepsilon > 0$ and $(c^* - \varepsilon, c^* + \varepsilon) \cap C = \emptyset$. Contradiction. Hence C is not finite.

Solution 10.51

Let $P \subset \mathbb{R}$ be a nonempty perfect set. Since P has a limit point (being not empty), P is infinite and is closed. Assume P is countable. Write $P = \{p_n; n \in \mathbb{N}\}$. Since p_0 is a limit point of P, $P \cap (p_0 - 1, p_0 + 1)$ is not empty and is infinite. Take a $p \in P \cap (p_0 - 1, p_0 + 1)$, with $p \neq p_0$. Then there exists an open interval I_1 which contains p, such that $\overline{I_0}$ does not contain p_0 and $\overline{I_0} \subset (p_0 - 1, p_0 + 1)$, where $\overline{I_0}$ denotes the closure of I_0. Since I_1 contains a limit point of P, it contains infinitely many points of P. In particular, it contains a point different from p_1. Since I_1 is open, it will contain an open interval I_2, which contains a point from P such that $\overline{I_2}$ does not contain p_1 and $\overline{I_2} \subset I_1$. By induction, we will construct a sequence of open intervals $\{I_n\}_{n \in \mathbb{N}} \subset (p_0 - 1, p_0 + 1)$ such that

1. $p_n \notin \overline{I_n}$,
2. $\overline{I_{n+1}} \subset I_n$,
3. $I_n \cap P \neq \emptyset$,

for any $n \in \mathbb{N}$. The sequence $\{\overline{I_n} \cap P\}$ is a decreasing sequence of bounded closed nonempty sets. Hence $I = \underset{n \in \mathbb{N}}{\cap} \overline{I_n} \cap P \neq \emptyset$. Let $p^* \in I$. Then $p^* \in \overline{I_n} \cap P$, for any $n \in \mathbb{N}$. Hence $p^* \neq p_n$, for any $n \in \mathbb{N}$. So $P \setminus \{p_n; n \in \mathbb{N}\} \neq \emptyset$. Contradiction.

Solution 10.52

Assume P not empty. Before we prove that P is perfect, let us prove that $C = A \setminus P$ is countable. Indeed, let $a \in C$, then there exists $r_1 < a < r_2$ with $r_1, r_2 \in \mathbb{Q}$ such that $(r_1, r_2) \cap A$ is countable. Hence
$$C \subset \bigcup \Big\{ (q_1, q_2) \cap A;\ q_1, q_2 \in \mathbb{Q} \text{ such that } (q_1, q_2) \cap A \text{ is countable} \Big\}.$$
Since a countable union of countable sets is countable, we conclude that C is countable. Next let us prove that P is perfect. Clearly any condensation point of A is also a limit point of A. The converse is not true in general. Indeed, if we take $A = \{1/n;\ n \geq 1\}$, then 0 is a limit of A but it is not a condensation point since A is countable. Let $a \in P'$, i.e., a is a limit point of P. Let $\varepsilon > 0$, then there exists $p \in (a - \varepsilon, a + \varepsilon) \cap P$, such that $p \neq a$. Since $(a - \varepsilon, a + \varepsilon)$ is open, there exists $\delta > 0$ such that $(p - \delta, p + \delta) \subset (a - \varepsilon, a + \varepsilon)$. Since $(p - \delta, p + \delta) \cap A$ is infinite not countable, then $(a - \varepsilon, a + \varepsilon) \cap A$ is infinite not countable, i.e., $a \in P$. So $P' \subset P$. Let $p \in P$. Then for any $\varepsilon > 0$, $(p - \varepsilon, p + \varepsilon) \cap A$ is infinite and not countable. Also $(p - \varepsilon, p + \varepsilon) \cap C$ is countable. Since
$$(p - \varepsilon, p + \varepsilon) \cap A = \Big((p - \varepsilon, p + \varepsilon) \cap P \Big) \cup \Big((p - \varepsilon, p + \varepsilon) \cap C \Big),$$
we conclude that $(p - \varepsilon, p + \varepsilon) \cap P$ is infinite and not countable. Hence p is a limit point of P, i.e., $p \in P'$. Therefore we have $P = P'$, or P is perfect.

Solution 10.53

Let P be the set of all condensed points of C. Set $F = C \setminus P$. Then $C = P \cup F$. We have $P \cap F = \emptyset$. According to Problem 10.52, P is perfect, and F is countable.

Chapter 11

Sequences and Series of Functions

Where is it proved that one obtains the derivative of an infinite series by taking derivative of each term?

Niels Henrik Abel (1802–1829)

- We say that a sequence of functions $\{f_n : D \to \mathbb{R}\}$ defined on a subset $D \subseteq \mathbb{R}$ *converges pointwise* on D if for each $x \in D$ the sequence of numbers $\{f_n(x)\}$ converge. If $\{f_n\}$ converges pointwise on D, then we define $f : D \to \mathbb{R}$ with $f(x) = \lim_{n \to \infty} f_n(x)$ for each $x \in D$. We denote this symbolically by $f_n \to f$ on D.

- We say that a sequence of functions $\{f_n\}$ defined on a subset $D \subseteq \mathbb{R}$ *converges uniformly* on D to a function f such that for every $\varepsilon > 0$ there is a number N such that

$$|f_n(x) - f(x)| < \varepsilon \quad \text{for all } x \in D, \ n \geq N.$$

We denote this type of convergence symbolically by $f_n \rightrightarrows f$ on D.

- If $\{f_n\}_0^\infty$ is a sequence of functions defined on D, the *series* $\sum_{n=0}^\infty f_n$ is said to *converge pointwise* (respectively, uniformly) on D if and only if the sequence $\{s_n\}_{n=0}^\infty$ of partial sums, given by

$$s_n(x) = \sum_{k=0}^n f_k(x),$$

converges pointwise (respectively, uniformly) on D.

- *Weierstrass M-Test*: Suppose that $\{f_n\}$ is a sequence of functions defined on D and $\{M_n\}$ is a sequence of nonnegative numbers such that
$$|f_n(x)| \leq M_n \quad \forall x \in D, \ \forall n \in \mathbb{N}.$$
If $\sum_{n=0}^{\infty} M_n$ converges, then $\sum_{n=0}^{\infty} f_n(x)$ converges uniformly on D.

- Assume that $f_n \rightrightarrows f$ uniformly on $[a,b]$ and that each f_n is integrable. Then, f is integrable and
$$\lim_{n \to \infty} \int_a^b f_n = \int_a^b f.$$

As a corollary to this theorem we obtain the following result:

- *Termwise Integration*: If each function $f_n(x)$ is continuous on a closed interval $[a,b]$ and if the series $\sum_{n=1}^{\infty} f_n(x)$ converges uniformly on $[a,b]$, then we have
$$\sum_{n=1}^{\infty} \int_a^b f_n(x)dx = \int_a^b \sum_{n=1}^{\infty} f_n(x)dx.$$

- Suppose that $\{f_n\}$ converges to f on the interval $[a,b]$. Suppose also that f'_n exists and is continuous on $[a,b]$, and the sequence $\{f'_n\}$ converges uniformly on $[a,b]$. Then
$$\lim_{n \to \infty} f'_n(x) = f'(x)$$
for each $x \in [a,b]$. As a corollary to this theorem we obtain the following result:

- *Termwise Differentiation*: If each function $f_n(x)$ has the derivative $f'_n(x)$ at any point $x \in (a,b)$, if the series $\sum_{n=1}^{\infty} f_n(x)$ converges to at least one point $k \in (a,b)$, and if $\sum_{n=1}^{\infty} f'_n(x)$ converges uniformly on (a,b) to a function $g(x)$, then $\sum_{n=1}^{\infty} f_n(x)$ converges uniformly on (a,b) and is differentiable at any point $x \in (a,b)$, whose derivative is equal to $g(x)$. In other words, termwise differentiation is possible, i.e.,
$$\Big(\sum_{n=1}^{\infty} f_n(x)\Big)' = \sum_{n=1}^{\infty} f'_n(x).$$

- A family of functions $\{f_n\} \in \mathcal{F}$ mapping a set $A \in \mathbb{R}^n$ into $\mathbb{R}^m$ is *equicontinuous* at a point $a \in A$ if for every $\varepsilon > 0$, $\exists \delta > 0$ such that
$$|f_n(x) - f_n(a)| < \varepsilon \text{ where } |x - a| < \delta \text{ and } f_n \in \mathcal{F}.$$

The family $\mathcal{F}$ is *equicontinuous on a set* A if it is equicontinuous at every point in A.

- *Arzelà–Ascoli Theorem*: Let A be a compact subset of $\mathbb{R}^n$ and $\mathcal{C}(A, \mathbb{R}^m)$ the space of continuous functions from A into $\mathbb{R}^m$. A subset $\mathcal{B}$ of $\mathcal{C}(A, \mathbb{R}^m)$ is compact if and only if it is closed, bounded, and equicontinuous.

CHAPTER 11. SEQUENCES AND SERIES OF FUNCTIONS

- *Weierstrass Approximation Theorem*: If f is a continuous function on a closed interval $[a, b]$, then there exists a sequence of polynomials that converge uniformly to $f(x)$ on $[a, b]$.

Problem 11.1 Prove that if a sequence $\{f_n\}$ of continuous functions on D converges uniformly to f on D, then f is continuous on D.

Problem 11.2 Prove that a sequence of functions $\{f_n\}$ defined on D is uniformly convergent on $U \subset D$ to $f : U \longrightarrow \mathbb{R}$ if and only if $\lim\limits_{n \to \infty} a_n = 0$, where

$$a_n := \sup\{|f_n(x) - f(x)| \; : \; x \in U\}, \quad n \in \mathbb{N}.$$

Problem 11.3 Consider the sequence $\{f_n\}$ defined by $f_n(x) = \dfrac{nx}{1+nx}$, for $x \geq 0$.

a) Find $f(x) = \lim\limits_{n \to \infty} f_n(x)$.

b) Show that for $a > 0$, $\{f_n\}$ converges uniformly to f on $[a, \infty)$.

c) Show that $\{f_n\}$ does not converge uniformly to f on $[0, \infty)$.

Problem 11.4 Consider the sequence $\{f_n\}$ defined by $f_n(x) = \dfrac{1}{1+x^n}$, for $x \in [0, 1]$.

a) Find $f(x) = \lim\limits_{n \to \infty} f_n(x)$.

b) Show that for $0 < a < 1$, $\{f_n\}$ converges uniformly to f on $[0, a]$.

c) Show that $\{f_n\}$ does not converge uniformly to f on $[0, 1]$.

Problem 11.5 Consider the sequence $\{f_n\}$ defined by $f_n(x) = \dfrac{nx}{e^{nx}}$, for $x \in [0, 2]$.

a) Show that $\lim\limits_{n \to \infty} f_n(x) = 0$ for $x \in (0, 2]$.

b) Show that the convergence is not uniform on $[0, 2]$.

Problem 11.6 Determine whether the sequence $\{f_n\}$ converges uniformly on D.

a) $f_n(x) = \dfrac{1}{1+(nx-1)^2}$ $\quad D = [0,1]$

b) $f_n(x) = nx^n(1-x)$ $\quad D = [0,1]$

c) $f_n(x) = \arctan\left(\dfrac{2x}{x^2+n^3}\right)$ $\quad D = \mathbb{R}$

Problem 11.7 Prove that if $\{f_n\}$ is a sequence of functions defined on A, then $\{f_n\}$ is uniformly convergent on D if and only if for $\varepsilon > 0$, there exists $N \in \mathbb{N}$ such that for any $m, n > N$
$$\sup_{x \in D} |f_n(x) - f_m(x)| < \varepsilon.$$
This is called the uniform Cauchy criterion.

Problem 11.8 Suppose that the sequence $\{f_n\}$ converges uniformly to f on the set D and that for each $n \in \mathbb{N}$, f_n is bounded on D. Prove that f is bounded on D.

Problem 11.9 Suppose a sequence of functions $\{f_n\}$ are defined as
$$f_n(x) = 2x + \frac{x}{n} \qquad x \in [0,1].$$

a) Find the limit function $f = \lim\limits_{n\to\infty} f_n$.

b) Is f continuous on [0,1]?

c) Does $[\lim\limits_{n\to\infty} f_n(x)]' = \lim\limits_{n\to\infty} f_n'(x)$ for $x \in [0,1]$?

d) Does $\displaystyle\int_0^1 \lim_{n\to\infty} f_n(x)\,dx = \lim_{n\to\infty} \int_0^1 f_n(x)\,dx$?

Problem 11.10 Give examples to illustrate that

a) the pointwise limit of continuous (respectively, differentiable) functions is not necessarily continuous (respectively, differentiable),

b) the pointwise limit of integrable functions is not necessarily integrable.

Problem 11.11 Give examples to illustrate that

a) there exist differentiable functions f_n and f such that $f_n \to f$ pointwise on $[0,1]$ but
$$\lim_{n\to\infty} f_n'(x) \neq \left(\lim_{n\to\infty} f_n(x)\right)' \quad \text{when } x = 1,$$

b) there exist continuous functions f_n and f such that $f_n \to f$ pointwise on $[0,1]$ but
$$\lim_{n\to\infty} \int_0^1 f_n(x)\,dx \neq \int_0^1 \left(\lim_{n\to\infty} f_n(x)\right) dx.$$

Problem 11.12 Suppose that $\{f_n\}$ is a sequence of functions defined on D and $\{M_n\}$ is a sequence of nonnegative numbers such that
$$|f_n(x)| \leq M_n \quad \forall x \in D, \ \forall n \in \mathbb{N}.$$
Show that if $\sum_{n=0}^{\infty} M_n$ converges, then $\sum_{n=0}^{\infty} f_n(x)$ converges uniformly on D. (This is called the Weierstrass M-test.)

Problem 11.13 Discuss the uniform convergence of the following series:

a) $\sum_{n=0}^{\infty} \dfrac{x^n}{n!}$ on $\mathbb{R}$

b) $\sum_{n=1}^{\infty} \dfrac{\sin(nx)}{\sqrt{n}}$ on $[0, 2\pi]$

c) $\sum_{n=1}^{\infty} \dfrac{\cos^2(nx)}{n^2}$ on $\mathbb{R}$

Problem 11.14 Let $\sum_{n=0}^{\infty} a_n x^n$ be a power series with radius of convergence r, where $0 < r \leq +\infty$. If $0 < t < r$, prove that the power series converges uniformly on $[-t, t]$.

Problem 11.15 Show that the function defined by $f(x) = \sum_{n=0}^{\infty} \left(\dfrac{x^n}{n!}\right)^2$ is continuous on $\mathbb{R}$.

Problem 11.16 Prove Dini's Theorem: Let $A \subseteq \mathbb{R}$ be a closed and bounded (and thus compact) set and $\{f_n\}$ be a sequence of continuous functions $f_n : A \longrightarrow \mathbb{R}$ such that

a) $f_n(x) \geq 0$ for any $x \in A$,

b) $f_n \longrightarrow f$ pointwise and f is continuous,

c) $f_{n+1}(x) \leq f_n(x)$ for any $x \in A$, and any $n \in \mathbb{N}$.

Prove that $\{f_n\}$ converges to f uniformly on A.

Problem 11.17 Give examples to illustrate that all of the hypotheses in Dini's Theorem (Problem 11.16) are essential.

Problem 11.18 Let $f_n : [1, 2] \longrightarrow \mathbb{R}$ be defined by

$$f_n(x) = \frac{x}{(1+x)^n}.$$

a) Show that $\sum_{n=1}^{\infty} f_n(x)$ converges for $x \in [1, 2]$.

b) Use Dini's Theorem to show that the convergence is uniform.

c) Does the following hold:

$$\int_1^2 \left(\sum_{n=1}^{\infty} f_n(x) \right) dx = \sum_{n=1}^{\infty} \int_1^2 f_n(x) dx?$$

Problem 11.19 A sequence of functions $\{f_n\}$ defined on a set A is said to be *equicontinuous* on A if for every $\varepsilon > 0$, there exists $\delta > 0$ such that

$$|f_n(x) - f_n(y)| < \varepsilon \text{ whenever } |x - y| < \delta$$

for $x, y \in A$ and $n \in \mathbb{N}$. Prove the following:

a) Any finite set of continuous functions on $[0,1]$ is equicontinuous.

b) If $\{f_n\}$ is a uniformly convergent sequence of continuous functions on $[0,1]$, then $\{f_n\}$ is equicontinuous.

CHAPTER 11. SEQUENCES AND SERIES OF FUNCTIONS

Problem 11.20 Show that there exists a continuous function defined on $\mathbb{R}$ that is nowhere differentiable by proving the following:

a) Let $g(x) = |x|$ if $x \in [-1, 1]$. Extend g to be periodic. Sketch g and the first few terms of the sum
$$f(x) = \sum_{n=1}^{\infty} \left(\frac{3}{4}\right)^n g(4^n x).$$

b) Use the Weierstrass M-test to show that f is continuous.

c) Prove that f is not differentiable at any point.

Problem 11.21 Let f_n be a function such that
$$f_n : (0, 1) \longrightarrow \mathbb{R}.$$

a) Prove that if $f_n \rightrightarrows f$ and $f_n' \rightrightarrows g$, then f is continuous on (0,1) and $f' = g$.

b) Describe how you would construct an example to show that uniform convergence of the derivatives is necessary.

Problem 11.22 Let $f_n : [0, 1] \longrightarrow \mathbb{R}$ be continuous such that $\{f_n\}$ are uniformly bounded on $[0, 1]$ and the derivatives f_n' exist and are uniformly bounded on $(0, 1)$. Prove that f_n has a uniformly convergent subsequence.

Problem 11.23 Let B be a bounded and equicontinuous subset of the set of continuous real-valued functions defined on $[a, b]$. Let $T : B \longrightarrow \mathbb{R}$ defined as
$$T(f) = \int_a^b f(x)dx.$$

Prove that there is a function $f_0 \in B$ at which the value of T is maximized.

Problem 11.24 Suppose $a, b, c,$ and d are constants chosen from an interval $[-K, K]$ and let $\Phi \subset (C[0, \pi], d)$ be a family of functions f of the form
$$f(x) = a \sin bx + c \cos dx \quad \text{where } 0 \leq x \leq \pi.$$
Metric d on $C[0, \pi]$ is given by $d(f, g) = \max\limits_{0 \leq x \leq \pi} |f(x) - g(x)|$.

a) Show that Φ is a compact subset of $C[0, \pi]$.

b) Show that for any continuous function g defined on $C[0, \pi]$ there exist values $a, b, c,$ and d in $[-K, K]$ such that
$$\max_{0 \leq x \leq \pi} |g(x) - (a \sin bx + c \cos dx)|$$
is minimum.
For an obvious reason $f \in \Phi$ is called *minimax approximation* of g.

Problem 11.25 (**Bernstein Polynomials**) The nth Bernstein polynomial of a continuous function $f : [0, 1] \to \mathbb{R}$ defined by
$$B_n(f)(x) = \sum_{k=0}^{n} f\left(\frac{k}{n}\right) \binom{n}{k} x^k (1-x)^{n-k}.$$

a) Show that B_n is linear, monotone map, and $B_n 1 = 1$ and $B_n x = x$.

b) Show that nth Bernstein polynomial for $f(x) = e^x$ is $B_n(x) = [1 + (e^{\frac{1}{n}} - 1)x]^n$.

c) Show that $B_n(e^x)$ converges uniformly to e^x on $[0, 1]$.

Problem 11.26

a) Show that for any function $f \in C[0, 1]$ and any number $\epsilon > 0$, there exists a polynomial p, all of whose coefficients are rational numbers, such that
$$\|p - f\| < \epsilon.$$

b) Show that $C[a, b]$ is separable.

Problem 11.27 Let f be a continuous function on $[a, b]$ and suppose that $\int_a^b f(x) x^n dx = 0$ for $n = 0, 1, 2, \ldots$. Prove that $f(x) = 0$ on $[a, b]$.

Solutions

Solution 11.1

It follows from the uniform convergence of $\{f_n\}$ that given $\varepsilon > 0$, $\exists N \in \mathbb{N}$ such that
$$|f_n(x) - f(x)| < \frac{1}{3}\varepsilon \quad \forall x \in D, \ n \geq N.$$

Let $c \in D$ be fixed. By the continuity of f_N at c, there is $\delta > 0$ such that
$$|f_N(x) - f_N(c)| < \frac{1}{3}\varepsilon \quad \text{whenever} \ |x - c| < \delta.$$

Thus
$$|f(x) - f(c)| \leq |f(x) - f_N(x)| + |f_N(x) - f_N(c)| + |f_N(c) - f(c)| < \varepsilon.$$

Solution 11.2

($\Rightarrow$) Suppose $f_n \rightrightarrows f$ on U. Then given $\varepsilon > 0$, there exists $N \in \mathbb{N}$ such that
$$|f_n(x) - f(x)| < \varepsilon \quad \forall n \geq N \ \forall x \in U.$$

Hence,
$$a_n := \sup\{|f_n(x) - f(x)| \ : \ x \in U, \ n \geq N\} < \varepsilon$$
and therefore $\lim_{n \to \infty} a_n = 0$.

($\Leftarrow$) Suppose $\lim_{n \to \infty} a_n = 0$. Then for large n and $\forall x \in U$
$$|f_n(x) - f(x)| \leq \sup\{|f_n(x) - f(x)| \ : \ x \in U\} < \varepsilon.$$

That is, $f_n \rightrightarrows f$ on U.

Solution 11.3

a) Since $f_n(0) = 0$ for all $n \in \mathbb{N}$, we get $f(0) = 0$. And for $x > 0$, we have
$$\lim_{n \to \infty} \frac{nx}{1 + nx} = \lim_{n \to \infty} \frac{1}{1 + 1/nx} = 1,$$
which yields $f(x) = 1$ for $x > 0$.

b) If $x \geq a$, then
$$\left|\frac{nx}{1 + nx} - 1\right| = \frac{1}{1 + nx} \leq \frac{1}{1 + na},$$
and thus we have
$$\lim_{n \to \infty} \frac{1}{1 + na} = 0,$$
which implies $\{f_n\}$ converges uniformly to f on $[a, \infty)$.

c) Let $n \geq 1$. If $0 < x < \dfrac{1}{n}$, then
$$\left|\dfrac{nx}{1+nx} - 1\right| = \dfrac{1}{1+nx} > \dfrac{1}{1+1} = \dfrac{1}{2},$$
which implies $\{f_n\}$ does not converge uniformly to f on $[0,\infty)$.

Solution 11.4

a) Since $f_n(1) = 1/2$ for all $n \in \mathbb{N}$, we get $f(1) = 1/2$. And for $0 \leq x < 1$, we have
$$\lim_{n\to\infty} \dfrac{1}{1+x^n} = 1,$$
which yields $f(x) = 1$ for $0 \leq x < 1$.

b) If $x \in [0,a]$, then
$$\left|\dfrac{1}{1+x^n} - 1\right| = \dfrac{x^n}{1+x^n} \leq \dfrac{a^n}{1+a^n}.$$
Furthermore, because $0 < a < 1$,
$$\lim_{n\to\infty} \dfrac{a^n}{1+a^n} = 0,$$
so $\{f_n\}$ converges uniformly to f on $[0,a]$.

c) Given $n \in \mathbb{N}$, let x be such that $\sqrt[n]{\dfrac{1}{2}} < x < 1$, then $\dfrac{1}{2} < x^n < 1$. Therefore
$$\left|\dfrac{1}{1+x^n} - 1\right| = \dfrac{x^n}{1+x^n} > \dfrac{\tfrac{1}{2}}{1+1} = \dfrac{1}{4},$$
which implies $\{f_n\}$ does not converge uniformly to f on $[0,1]$.

Solution 11.5

a) Since
$$\dfrac{f_{n+1}}{f_n} = \dfrac{(n+1)xe^{nx}}{nxe^{(n+1)x}} = \dfrac{(n+1)}{n}e^{-x},$$
$\lim\limits_{n\to\infty} \dfrac{f_{n+1}}{f_n} = e^{-x} < 1$. Now we use Problem 3.16 to show that $\lim\limits_{n\to\infty} f_n = 0$ for $x \in (0,2]$.

b) Let us find the maximum of f_n in $[0,2]$. Since
$$f'_n(x) = \dfrac{ne^{nx} - ne^{nx}nx}{(e^{nx})^2} = 0 \implies e^{nx}(n - n^2 x) = 0$$
$$\implies n^2 x = n$$
$$\implies x = \dfrac{1}{n},$$

it is easy to check that $1/n$ is the maximum with $f_n\left(\dfrac{1}{n}\right) = \dfrac{1}{e}$. Since

$$\lim_{n\to\infty} \sup\{|f_n(x) - 0| \ : \ x \in [0,2]\} = \lim_{n\to\infty} f_n\left(\dfrac{1}{n}\right) = \dfrac{1}{e},$$

the convergence is not uniform on $[0,2]$.

Solution 11.6

a) Note
$$\lim_{n\to\infty} f_n(x) = \lim_{n\to\infty} \dfrac{1}{1+(nx-1)^2} = \begin{cases} 0 & x \in (0,1], \\ \dfrac{1}{2} & x = 0. \end{cases}$$

Notice that all f_n's are continuous while the limit function is not continuous, which implies that convergence is not uniform.

b) First notice that
$$\lim_{n\to\infty} f_n(x) = \lim_{n\to\infty} nx^n(1-x) = 0.$$

The previous argument will not work in this case. So let us find the maximum value of f_n, for $n \geq 1$. Let $n \geq 2$. We have

$$f_n'(x) = n^2 x^{n-1}(1-x) - nx^n = nx^{n-1}\Big(n - (n+1)x\Big).$$

Hence $f_n'(x) = 0$ iff $x = 0$ or $x = \dfrac{n}{n+1}$. It is easy to check that the maximum value of $f_n(x)$ is given by
$$\sup\{|f_n(x) - 0| \ : \ x \in [0,1]\} = f_n\left(\dfrac{n}{n+1}\right).$$

However,
$$\lim_{n\to\infty} f_n\left(\dfrac{n}{n+1}\right) = \lim_{n\to\infty} \left(\dfrac{n}{n+1}\right)^{n+1} = \dfrac{1}{e} \neq 0.$$

Thus convergence is not uniform.

c) We have
$$\lim_{n\to\infty} f_n(x) = \lim_{n\to\infty} \arctan\left(\dfrac{2x}{x^2+n^3}\right) = 0.$$

In order to find the maximum value of f_n, let us compute its derivative
$$f_n'(x) = \dfrac{2n^3 - 2x^2}{(x^2+n^3)^2 + 4x^2}.$$

Hence $f_n'(x) = 0$ iff $x = \pm n\sqrt{n}$. It is easy to check that the maximum value of $|f_n(x)|$ is given by
$$\sup\{|f_n(x) - 0| \ : \ x \in \mathbb{R}\} = f_n(n\sqrt{n}) = \arctan\left(\dfrac{1}{n\sqrt{n}}\right).$$

Since $\lim_{n\to\infty} \sup\{|f_n(x) - 0| \ : \ x \in \mathbb{R}\} = 0$, $\{f_n\}$ converges uniformly to 0 on $\mathbb{R}$.

Solution 11.7

($\Rightarrow$) Suppose $f_n \rightrightarrows f$, then given $\varepsilon > 0$, we can find an integer N such that

$$m \geq N \implies |f_m(x) - f(x)| < \frac{\varepsilon}{2}$$

for all $x \in D$. If $m, n \geq N$, then

$$|f_m(x) - f_n(x)| \leq |f_n(x) - f(x)| + |f(x) - f_n(x)| < \frac{\varepsilon}{2} + \frac{\varepsilon}{2} = \varepsilon$$

for any $x \in D$.

($\Leftarrow$) If given some $\varepsilon > 0$, we can find an N such that for any $m, n \geq N$ we have

$$\sup_{x \in D} |f_n(x) - f_m(x)| < \varepsilon.$$

Hence for any $x \in D$, $\{f_n(x)\}$ is a Cauchy sequence in $\mathbb{R}$ which implies $f_n(x)$ converges to $f(x)$. Let us prove that $\{f_n\}$ converges uniformly to f on D. Indeed, let $\varepsilon > 0$. By the uniform Cauchy criterion, there exists $N \in \mathbb{N}$ such that for all $n, m \geq N$, we have $\sup_{x \in D} |f_n(x) - f_m(x)| \leq \varepsilon/2$. Fix $n \geq N$. Then

$$|f_n(x) - f(x)| = \lim_{m \to \infty} |f_n(x) - f_m(x)| \leq \frac{\varepsilon}{2}$$

for any $x \in D$. Hence

$$\sup_{x \in D} |f_n(x) - f(x)| \leq \frac{\varepsilon}{2} < \varepsilon.$$

This obviously implies $f_n \rightrightarrows f$ on D.

Solution 11.8

Since $\{f_n\}$ converges to f uniformly on D, there exists an N such that $n \geq N$ implies that $|f_n(x) - f(x)| < 1$ for all $x \in D$. Fix $n_0 \geq N$, since f_{n_0} is bounded on D, there exists a constant k such that $|f_{n_0}(x)| < k$ for all $x \in D$. Thus for all $x \in D$ we have that

$$|f(x)| \leq |f(x) - f_{n_0}(x)| + |f_{n_0}(x)| \leq 1 + k.$$

Thus f is bounded on D.

Solution 11.9

a) $f(x) = 2x$ for $x \in [0, 1]$.

b) $2x$ is continuous for all $x \in [0, 1]$.

c) Note that $\lim_{n \to \infty} f'_n(x) = \lim_{n \to \infty} 2 + \frac{1}{n} = 2$. Therefore $[\lim_{n \to \infty} f_n(x)]' = f'(x) = 2$.

d) Note that $\lim\limits_{n\to\infty} \int_0^1 f_n(x)dx = \lim\limits_{n\to\infty} \int_0^1 2x + \dfrac{x}{n} dx = \lim\limits_{n\to\infty} 1 + \dfrac{1}{2n} = 1$ and $\int_0^1 f(x)dx = 1$.
Therefore
$$\int_0^1 \lim\limits_{n\to\infty} f_n(x)dx = \int_0^1 f(x)dx = 1.$$

Solution 11.10

a) Let $f_n(x) = x^n$, for $x \in [0,1]$, as in the following figure:

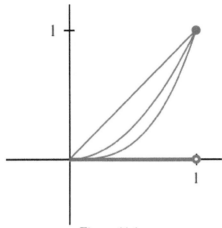

Figure 11.1

Then $\{f_n\}$ converges pointwise to
$$f(x) = \begin{cases} 0 & 0 \leq x < 1, \\ 1 & x = 1. \end{cases}$$

Each f_n is continuous and differentiable on $[0,1]$, but f is neither continuous nor differentiable at $x = 1$.

b) The pointwise limit of integrable functions is not necessarily integrable. Indeed, let
$$f_n(x) = \begin{cases} 1 & \text{if } x = \dfrac{p}{m} \in \mathbb{Q},\ m \leq n, \text{ when written in reduced form,} \\ 0 & \text{otherwise,} \end{cases}$$

for $n \in \mathbb{N}$. Then $\{f_n\}$ converges pointwise to
$$f(x) = \begin{cases} 1 & x \in \mathbb{Q}, \\ 0 & \text{otherwise.} \end{cases}$$

Since each of the f_n has only a finite number of nonzero points, it is integrable on $[0,1]$ with integral zero. However, f is not integrable on $[0,1]$ as can be seen from Problem 7.2.

Solution 11.11

a) Let $f_n(x) = x^n/n$, for $n \geq 1$, and $f(x) = 0$. Then $f_n \to f$ pointwise on [0,1]. Each f_n is differentiable and $f'_n(x) = x^{n-1}$. Thus, for $x \in [0,1)$,

$$\lim_{n\to\infty} f'_n(x) = \lim_{n\to\infty} x^{n-1} = 0.$$

When $x = 1$, however,
$$\lim_{n\to\infty} f'_n(1) = \lim_{n\to\infty} 1^{n-1} = 1.$$

Thus, for $x = 1$,
$$\lim_{n\to\infty} f'_n(1) = 1 \neq 0 = f'(1).$$

b) For $x \in [0,1]$, set $f_1(x) = 1$ and for $n \geq 2$

$$f_n(x) = \begin{cases} n^2 x & 0 \leq x \leq 1/n, \\ 2n - n^2 x & 1/n < x \leq 2/n, \\ 0 & 2/n < x \leq 1, \end{cases}$$

as depicted in the following figure:

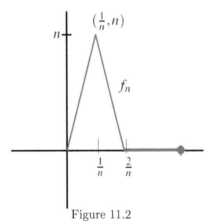

Figure 11.2

Since f_n encloses an area of a triangle with base $2/n$ and altitude n, we have

$$\int_0^1 f_n(x)\,dx = 1 \quad \text{for all } n > 2$$

which can also be computed directly. Therefore,

$$\lim_{n\to\infty} \int_0^1 f_n(x)\,dx = 1.$$

However, $f_n \to f$ where $f(x) = 0$ for $x \in [0,1]$. This is because for $x = 0$, $f_n(0) = 0$ for every n, and for any $x \in (0,1]$, we also have $f_n(x) = 0$ for large enough n. Therefore,

$$\int_0^1 \left(\lim_{n \to \infty} f_n(x)\right) dx = 0.$$

Thus we have the result that

$$\lim_{n \to \infty} \int_0^1 f_n(x) dx = 1 \neq 0 = \int_0^1 \left(\lim_{n \to \infty} f_n(x)\right) dx.$$

Solution 11.12

Let $s_n(x) = \sum_{k=0}^{n} f_k(x)$ be the nth partial sum. Since $\sum_{n=0}^{\infty} M_n$ converges, for all $\varepsilon > 0$, $\exists N$ such that if $n \geq m \geq N$, then

$$M_m + M_{m+1} + \cdots + M_n < \varepsilon.$$

Thus if $n \geq m \geq N$, we have

$$\begin{aligned} |s_n(x) - s_m(x)| &= |f_{m+1}(x) + \cdots + f_n(x)| \\ &\leq |f_{m+1}(x)| + \cdots + |f_n(x)| \\ &\leq M_{m+1} + \cdots + M_n < \varepsilon \end{aligned}$$

for all $x \in D$. It follows from Problem 11.7 that $\{s_n\}$ converges uniformly on D. Hence $\sum_{n=0}^{\infty} f_n$ also converges uniformly on D.

Solution 11.13

a) For any $n \geq 1$, we have

$$\sup_{x \in \mathbb{R}} |s_n(x) - s_{n-1}(x)| = \sup_{x \in \mathbb{R}} |f_n(x)| \geq |f_n(n)| = \frac{n^n}{n!} \geq 1.$$

The Cauchy criterion is not satisfied and thus, convergence is not uniform.

b) First note that we do have pointwise convergence. Next notice that

$$\frac{2}{\pi} x \leq \sin(x), \text{ for any } x \in \left[0, \frac{\pi}{4}\right].$$

Let $n \geq 10$, and $h \in \mathbb{N}$ such that $2n \leq n + h < n\sqrt{n}\pi/4$. Then for any $k \in \mathbb{N}$ with $n \leq k \leq n + h$, we have $k/n\sqrt{n} < \pi/4$. Hence

$$\frac{2}{\pi} \frac{k}{n\sqrt{n}} \leq \sin\left(k \frac{1}{n\sqrt{n}}\right),$$

which implies
$$\sum_{k=n}^{n+h} \frac{1}{\sqrt{k}} \frac{2}{\pi} \frac{k}{n\sqrt{n}} \leq \sum_{k=n}^{n+h} \frac{1}{\sqrt{k}} \sin\left(k\frac{1}{n\sqrt{n}}\right).$$

But
$$\sum_{k=n}^{n+h} \frac{1}{\sqrt{k}} \frac{2}{\pi} \frac{k}{n\sqrt{n}} = \sum_{k=n}^{n+h} \frac{2}{\pi} \frac{\sqrt{k}}{n\sqrt{n}} \geq \sum_{k=n}^{n+h} \frac{2}{\pi} \frac{\sqrt{n}}{n\sqrt{n}} = \frac{2}{\pi} \frac{h}{n} \geq \frac{2}{\pi}.$$

This obviously shows that
$$\sup_{x\in[0,2\pi]} \left|\sum_{k=n}^{n+h} \frac{1}{\sqrt{k}} \sin(kx)\right| \geq \frac{2}{\pi}$$

for any $n \geq 10$ and $h \in \mathbb{N}$ such that $2n \leq n+h < n\sqrt{n}\pi/4$. Therefore the convergence will not be uniform on $[0, 2\pi]$.

c) Let $M_n = 1/n^2$, for $n \geq 1$, then
$$\left|\frac{\cos^2(nx)}{n^2}\right| \leq M_n,$$

since $|\cos(nx)| \leq 1$. Because $\sum_{n=1}^{\infty} \frac{1}{n^2}$ is a p-series with $p = 2 > 1$, it converges. Hence, by the Weierstrass M-test convergence is uniform (see Problem 11.12).

Solution 11.14

Let $x \in [-t, t]$, then $|a_n x^n| \leq |a_n| t^n$. Since $0 < t < r$, the series $\sum_{n=0}^{\infty} |a_n| t^n$ is convergent. Thus by the Weierstrass M-test, $\sum_{n=0}^{\infty} a_n x^n$ converges uniformly on $[-t, t]$.

Solution 11.15

Let $t \in \mathbb{R}$. Let us first show that $\sum_{n=0}^{\infty} \left(\frac{x^n}{n!}\right)^2$ converges uniformly on $[-t, t]$. Set
$$M_n = \left(\frac{t^n}{n!}\right)^2, \quad \text{for } n \geq 1.$$

Since
$$\lim_{n\to\infty} \frac{M_{n+1}}{M_n} = \lim_{n\to\infty} \left(\frac{t^{n+1}}{(n+1)!}\right)^2 \cdot \left(\frac{n!}{t^n}\right)^2 = \lim_{n\to\infty} \left(\frac{t}{n+1}\right)^2 = 0 < 1,$$

the ratio test forces $\sum_{n=0}^{\infty} M_n$ to be convergent and by the Weierstrass M-test, $\sum_{n=0}^{\infty} \left(\frac{x^n}{n!}\right)^2$ converges uniformly on $[-t, t]$. Uniform convergence will guarantee that $f(x)$ is continuous because each of the partial sums is continuous on $[-t, t]$. Since t was arbitrary, we have continuity of $f(x)$ on $\mathbb{R}$.

Solution 11.16

Let $\{g_n\} = \{f_n - f\}$. It is sufficient to show that $g_n \rightrightarrows 0$ on A. Given $\varepsilon > 0$, we want to find N such that $|g_n(x)| < \varepsilon$ for all $n \geq N$ and $x \in A$. By hypothesis b), we know that $\{g_n\}$ converges to 0 pointwise on A. That is, for $x \in A$ there exists $N_x = N(\varepsilon, x)$ such that for $n \geq N_x$, $0 \leq g_n(x) \leq \varepsilon/2$. By the continuity of g_{N_x} there is a neighborhood $U(x, N_x)$ such that

$$|g_{N_x}(y) - g_{N_x}(x)| < \frac{\varepsilon}{2} \quad \text{for any } y \in U(x, N_x).$$

The neighborhoods $U(x_i, N_{x_i})$ form a cover for the compact set A. That is, there are finitely many points $x_1, x_2, \ldots, x_n \in A$ such that

$$A \subset U(x_1, N_{x_1}) \cup U(x_2, N_{x_2}) \cup \cdots \cup U(x_n, N_{x_n}).$$

Let $N = \max(N_{x_1}, N_{x_2}, \ldots, N_{x_n})$. For any $x \in A$, there exists x_i such that $x \in N_{x_i}$. Hence for any $n \geq N$ we have

$$0 \leq g_n(x) \leq g_{N_{x_i}}(x) \leq g_{N_{x_i}}(x) - g_{N_{x_i}}(x_i) + g_{N_{x_i}}(x_i) \leq \frac{\varepsilon}{2} + \frac{\varepsilon}{2} = \varepsilon.$$

Therefore $|g_n(x)| < \varepsilon$ for $n > N$, $x \in A$. This completes our proof.

Solution 11.17

(1) Let

$$f_n(x) = \frac{1}{1 + nx} \quad x \in (0, 1), \ n = 1, 2, \ldots.$$

It is the case that

$$f_{n+1}(x) \leq f_n(x).$$

However, using Problem 11.2, we see

$$a_n = \sup_{x \in (0,1)} \left| \frac{1}{1 + nx} - 0 \right| = 1,$$

so convergence is not uniform, because $A = (0, 1)$ is not closed, hence not compact.

(2) The assumption that $\{f_n\}$ is a monotone sequence is also necessary. If we consider the function defined in Problem 11.11 (b), then $f_n \to f$ pointwise on $[0,1]$. But the sequence $\{f_n\}$ is not monotonic and convergence is not uniform.

(3) Continuity of each f_n cannot be omitted. For instance, let

$$f_n(x) = \begin{cases} 1 & x \in (0, \frac{1}{n}), \\ 0 & x = 0 \text{ or } x \in [\frac{1}{n}, 1]. \end{cases}$$

Then each f_n is not continuous. However, they form a monotonic sequence converging pointwise to zero on $[0,1]$, but again the convergence is not uniform.

(4) Finally, the continuity of the limit function is also needed. The sequence $f_n(x) = x^n$ for $x \in [0, 1]$ defined in Problem 11.10 (a) has a discontinuous limit $f(x)$ and $f_n(x)$ fails to converge uniformly on $[0,1]$.

Solution 11.18

a) First observe that
$$\sum_{n=1}^{\infty} \frac{x}{(1+x)^n} = x \sum_{n=1}^{\infty} \frac{1}{(1+x)^n}$$
and when $|1/(1+x)| < 1$ or equivalently when $|1+x| > 1$ we have that
$$\sum_{n=1}^{\infty} \frac{x}{(1+x)^n} = \frac{x}{1+x} \cdot \frac{1}{1 - \frac{1}{1+x}} = 1$$
so in particular $\sum_{n=1}^{\infty} f_n(x)$ is convergent for $x \in [1,2]$.

b) $A = [1,2]$ is compact and $f_n(x) \to 0$ pointwise. Clearly, if $k \geq \ell$ holds, then $f_\ell(x) \geq f_k(x)$. All the hypotheses of Dini's theorem are satisfied and thus convergence is uniform.

c) Since the convergence is uniform we can interchange the integral and the summation. Thus the equality holds.

Solution 11.19

a) Consider $\{f_i : 1 \leq i \leq n\}$ where each $f_i : [0,1] \to \mathbb{R}$ is continuous. Since $[0,1]$ is compact, each f_i is uniformly continuous. Let $\varepsilon > 0$. Since f_i is uniformly continuous, there exists $\delta_i > 0$ such that if $|x - y| < \delta_i$, then $|f_i(x) - f_i(y)| < \varepsilon$ for $1 \leq i \leq n$. Now let $\delta = \min\{\delta_1, \delta_2, \ldots, \delta_n\}$, then for any $x, y \in [0,1]$ such that $|x - y| < \delta$, we have
$$|f_i(x) - f_i(y)| < \varepsilon, \quad \text{for } 1 \leq i \leq n.$$
Hence $\{f_i\}$ is equicontinuous.

b) Let $\varepsilon > 0$. Since $\{f_n\}$ is uniformly convergent, there exists $N \in \mathbb{N}$ such that for any $n, m \geq N$ we have
$$\sup_{x \in [0,1]} |f_n(x) - f_m(x)| < \frac{\varepsilon}{3}.$$
In particular, we have for any $n \geq N$
$$\sup_{x \in [0,1]} |f_n(x) - f_N(x)| < \frac{\varepsilon}{3},$$
for any $n \geq N$. The first part shows that the family $\{f_1, f_2, \ldots, f_N\}$ is equicontinuous. Hence there exists $\delta > 0$ such that
$$|f_i(x) - f_i(y)| < \frac{\varepsilon}{3}$$
for any $i \leq N$ and $x, y \in [0,1]$ such that $|x - y| < \delta$. Let $n \geq N$, then we have
$$|f_n(x) - f_n(y)| \leq |f_n(x) - f_N(x)| + |f_N(x) - f_N(y)| + |f_N(y) - f_n(y)| < \varepsilon$$

for any $x, y \in [0, 1]$ such that $|x - y| < \delta$. Hence for any $n \in \mathbb{N}$ and any $x, y \in [0, 1]$ with $|x - y| < \delta$, we have
$$|f_n(x) - f_n(y)| < \varepsilon.$$
Therefore $\{f_n\}$ is equicontinuous.

Solution 11.20

a) Extend $g(x)$ to all of $\mathbb{R}$ by requiring $g(x) = g(x+2)$ for all x. This is depicted in the following figure:

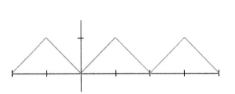
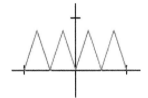

Figure 11.3

b) Now, given $x, y \in \mathbb{R}$, we have
$$|g(x) - g(y)| = \Big||x| - |y|\Big| \leq |x - y|.$$

By translation, this also applies to any pair of points that are no more than two apart. On the other hand, if $|x - y| > 2$, then
$$|g(x) - g(y)| \leq |g(x)| + |g(y)| \leq 1 + 1 = 2 < |x - y|.$$

Hence, $|g(x) - g(y)| \leq |x - y|$ for all x and y, and so it follows that g is continuous on $\mathbb{R}$.

For each integer $n \geq 0$, let $g_n(x) = (3/4)^n g(4^n x)$. For instance,
$$\begin{aligned} g_0(x) &= g(x) \\ g_1(x) &= \tfrac{3}{4} g(4x) \\ g_2(x) &= \tfrac{9}{16} g(16x) \\ &\vdots \end{aligned}$$

Notice that g_n oscillates four times as fast as g_{n-1} and at $3/4$ the height of g_{n-1}. Now $f(x) = \sum_{n=0}^{\infty} g_n(x)$ is defined on $\mathbb{R}$. Moreover, for all $x \in \mathbb{R}$, we have that
$$|g_n(x)| \leq \left(\frac{3}{4}\right)^n \qquad \forall n \in \mathbb{N}.$$

Thus we can apply the Weierstrass M-test to conclude that $\sum_{n=0}^{\infty} g_n(x)$ converges uniformly on $\mathbb{R}$. Since each g_n is continuous on $\mathbb{R}$, then $f(x)$ is continuous on $\mathbb{R}$.

c) To see that f is nowhere differentiable, we fix $x \in \mathbb{R}$ and let $h_m = \pm \frac{4^{-m}}{2}$ (note that $4^m |h_m| = \frac{1}{2}$). Next, we claim that
$$\left| \frac{f(x+h_m) - f(x)}{h_m} \right|$$
are bounded below by a sequence that diverges to $+\infty$ as $m \to \infty$. In order to accomplish this, we first observe that

$$f(x + h_m) - f(x) = \sum_{n=0}^{\infty} \left(\frac{3}{4}\right)^n [g(4^n x + 4^n h_m) - g(4^n x)].$$

Now we need to examine $g(4^n x + 4^n h_m) - g(4^n x)$ for each n.

Case 1: If $n > m$, then $4^n h_m = \pm 4^{n-m}/2$ is an even integer. Since $g(t) = g(t+2)$ for all t, it follows that $g(4^n x + 4^n h_m) - g(4^n x) = 0$.

Case 2: If $n < m$, then since $|g(r) - g(s)| \leq |r - s|$ for all $r, s \in \mathbb{R}$, we have that
$$|g(4^m x + 4^m h_m) - g(4^m x)| \leq 4^m |h_m|.$$

Case 3: If $n = m$, then since there is no integer between $4^m x$ and $4^m(x + h_m)$, the graph of g between these points is a straight line of slope ± 1. Thus
$$|g(4^m x + 4^m h_m) - g(4^m x)| = 4^m |h_m|.$$

Combining all of these three cases together, we obtain

$$\left| \frac{f(x+h_m) - f(x)}{h_m} \right| = \left| \sum_{n=0}^{m} \left(\frac{3}{4}\right)^n \frac{g(4^n x + 4^n h_m) - g(4^n x)}{h_m} \right|$$

$$\geq 3^m - \sum_{n=0}^{m-1} \left(\frac{3}{4}\right)^n \frac{4^n |h_m|}{|h_m|}$$

$$= 3^m - \sum_{n=0}^{m-1} 3^n$$

$$= 3^m - \frac{1 - 3^m}{1 - 3} = \frac{1}{2}(3^m + 1).$$

Since $(3^m + 1)/2$ diverges to $+\infty$ as $m \to \infty$, we have shown that f is not differentiable at x. Since x was arbitrary, f is nowhere differentiable.

Solution 11.21

1. Let $x_0 \in (0,1)$. We have
$$f_n(x) = f_n(x_0) + \int_{x_0}^{x} f_n'(u)du$$
for $x \in (0,1)$. If we take the limit as $n \to \infty$, we will get
$$\begin{aligned} f(x) - f(x_0) = \lim_{n\to\infty} (f_n(x) - f_n(x_0)) &= \lim_{n\to\infty} \int_{x_0}^{x} f_n'(u)du \\ &= \int_{x_0}^{x} g(u)du \end{aligned}$$
because uniform convergence allows for interchanging the limit with the integral. The equation $f(x) - f(x_0) = \int_{x_0}^{x} g(u)du$ implies $f'(x) = g(x)$, for any $x \in (0,1)$.

2. Consider the sequence $\{f_n\}$ defined by
$$f_n(x) = \frac{x}{1+nx^2}.$$
Then $\{f_n\}$ converges uniformly to 0 on $\mathbb{R}$. Indeed let $\varepsilon > 0$. Choose $N \in \mathbb{N}$ such that $1/\varepsilon^2 < N$. Then for any $n \geq N$ and $x \in \mathbb{R}$ we have

- Case 1: if $|x| < \varepsilon$, then
$$|f_n(x)| = \frac{|x|}{1+nx^2} \leq |x| < \varepsilon.$$

- Case 2: if $|x| \geq \varepsilon$, then
$$|f_n(x)| = \frac{|x|}{1+nx^2} \leq \frac{|x|}{nx^2} \leq \frac{1}{\varepsilon n} < \varepsilon,$$
since $n \geq N$.

Hence
$$\sup_{x\in\mathbb{R}} |f_n(x)| < \varepsilon, \text{ for any } n \geq N.$$
On the other hand, we have
$$f_n'(x) = \frac{1-nx^2}{(1+nx^2)^2}.$$
Hence $\{f_n'\}$ converges pointwise to g where
$$g(x) = \begin{cases} 1 & \text{if } x = 0, \\ 0 & \text{if } x \neq 0. \end{cases}$$
Notice that $f_n \rightrightarrows f$ uniformly, but $f_n' \to g$ pointwise and f is differentiable but $f' \neq g$.

Solution 11.22

This is known as the Arzelà–Ascoli Theorem in its simplest form, if we omit the uniform boundedness of the derivatives and replace it by the equicontinuity of the functions. Indeed since f'_n are uniformly bounded on $(0,1)$, there exists $M > 0$ such that $|f'_n(x)| \leq M$, for any $x \in (0,1)$. Using the Mean Value Theorem, we get

$$|f_n(x) - f_n(y)| \leq M|x - y|, \text{ for any } x, y \in [0, 1], \text{ and } n \in \mathbb{N}.$$

So if $\varepsilon > 0$ is given, set $\delta = \varepsilon/M$. Then $|f_n(x) - f_n(y)| < \varepsilon$, for any $n \in \mathbb{N}$, provided $|x - y| < \delta$. So $\{f_n\}$ is equicontinuous. Let us prove that $\{f_n\}$ has a subsequence which converges uniformly. Set $\{r_n\} = [0,1] \cap \mathbb{Q}$. Since $\{f_n\}$ is uniformly bounded, so $\{f_n(x)\}$ is bounded, for any $x \in [0,1]$. Therefore the Bolzano–Weierstrass Theorem shows that there exists a subsequence $\{f_{n_1}\}$ of $\{f_n\}$ such that $\{f_{n_1}(r_1)\}$ is convergent. By induction, we can construct a sequence of subsequences $\{f_{n_i}\}$ such that

(1) $\{f_{n_{i+1}}\}$ is a subsequence of $\{f_{n_i}\}$;

(2) $\{f_{n_i}(r_k)\}$, $k = 1, \ldots, i$ is convergent.

Consider the subsequence $\{f_{n_n}\}$. We have

(1) $\{f_{n_n}\}$ is a subsequence of $\{f_n\}$;

(2) $\{f_{n_n}(r_k)\}$, for any $k \in \mathbb{N}$, is convergent.

Let $\varepsilon > 0$. Since $\{f_n\}$ is equicontinuous, then there exists $\delta > 0$ such that $|f_n(x) - f_n(y)| < \varepsilon/3$, for any $n \in \mathbb{N}$, provided $|x - y| < \delta$. Since $[0,1]$ is compact, there exists $\{r_{n_1}, \ldots, r_{n_K}\}$ such that for any $x \in [0,1]$ there exists $1 \leq i \leq K$ such that $|x - r_{n_i}| < \delta$. Since $\{f_{n_n}(r_k)\}$, for any $k \in \mathbb{N}$, is convergent, there exists $N \in \mathbb{N}$ such that for any $n, m \geq N$,

$$|f_{n_n}(r_{n_i}) - f_{m_m}(r_{n_i})| < \frac{\varepsilon}{3}$$

for any $i = 1, \ldots, K$. Let $x \in [0,1]$, then there exists $1 \leq i \leq K$ such that $|x - r_{n_i}| < \delta$. Hence

$$|f_{n_n}(x) - f_{m_m}(x)| \leq |f_{n_n}(x) - f_{n_n}(r_{n_i})| + |f_{n_n}(r_{n_i}) - f_{m_m}(r_{n_i})| + |f_{m_m}(r_{n_i}) - f_{m_m}(x)| < \varepsilon.$$

So the subsequence $\{f_{n_n}\}$ satisfies the uniform Cauchy criteria which imply that $\{f_{n_n}\}$ is uniformly convergent.

Solution 11.23

B is closed, bounded, and equicontinuous. Therefore, by the Arzelà–Ascoli Theorem B is compact. Next we claim that $T : B \to \mathbb{R}$ defined as $T(f) = \int_a^b f(x)dx$ is continuous. Let $\{f_k\}$ be a sequence of functions in B converging to some f uniformly. For $\varepsilon > 0$, we can choose N such that $k \geq N$ implies that $|f_k(x) - f(x)| < \dfrac{\varepsilon}{b-a}$. Then

$$\left| \int_a^b f_k(x)dx - \int_a^b f(x)dx \right| = \left| \int_a^b f_k(x) - f(x)dx \right|$$
$$\leq \int_a^b |f_k(x) - f(x)| \, dx$$
$$\leq \frac{\varepsilon}{b-a}(b-a) = \varepsilon.$$

CHAPTER 11. SEQUENCES AND SERIES OF FUNCTIONS

T is continuous on a compact set B, hence $T(B)$ is also compact. Since compact subsets of $\mathbb{R}$ have a maximum, there exists $f_0 \in B$ such that

$$T(f_0) = \max\{T(f); f \in B\}.$$

Solution 11.24

a) The family Φ is uniformly bounded and equicontinuous, since

$$|\hat{f}(x)| \leq |ab| + |cd| \leq 2K^2 \text{ and } |f(x)| \leq |a| + |c| \leq 2K$$

for all $x \in [0, \pi]$ and any $f \in \Phi$. Furthermore Φ is a closed subset of $C[0, \pi]$, by the Arzelà–Ascoli Theorem we conclude that Φ is compact in $C[0, \pi]$.

b) Take $g \in C[0, \pi]$, consider the distance $d(g, \Phi) = \inf\{d(g, h) : h \in \Phi\}$. Since Φ is compact this infimum is attained. In other words there exist values of $a, b, c,$ and d in $[-K, K]$ such that

$$d(f, g) = \max_{0 \leq x \leq \pi} |g(x) - (a\sin bx + c\cos dx)|$$

is a minimum.

Solution 11.25

a) It is clear that the map B_n is linear, since $B_n(f + g) = B_n f + B_n g$ and $B_n(\alpha f) = \alpha B_n f$.

Notice that when $f \geq 0$, then $B_n f$ is also positive. In particular, $|f| \leq g$ means $-g \leq f \leq g$ and hence $-B_n g \leq B_n f \leq B_n g$. This also proves that $|B_n f| < B_n g$ if $|f| < g$ (since it is straightforward to show that $f \geq g \Rightarrow B_n f \geq B_n g$).

b) $B_n(1) = \sum_{k=0}^n 1\binom{n}{k} x^k (1-x)^{n-k} = 1$ (From the Binomial Theorem). Next notice that

$$\frac{k}{n}\binom{n}{k} = \frac{k}{n}\frac{n!}{k!(n-k)!} = \frac{(n-1)!}{(k-1)!(n-k)!} = \binom{n-1}{k-1}.$$

Using the Binomial Theorem again, we have

$$B_n(x) = \sum_{k=0}^n \frac{k}{n}\binom{n}{k} x^k (1-x)^{n-k}$$

$$= x \sum_{k=0}^n \binom{n-1}{k-1} x^{k-1} (1-x)^{n-k}$$

$$= x(x + (1-x))^{n-1} = x.$$

c)
$$B_1(x) = f(0)(1-x) + f(1)x = (1-x) + ex = 1 + (e-1)x.$$
$$B_2(x) = f(0)(1-x)^2 + 2f(\frac{1}{2})x(1-x) + f(1)x^2 = (10x)^2 + 2e^{1/2}x(1-x)ex^2$$
$$= \left((1-x) + e^{1/2}x\right)^2 = \left(1 + (e^{1/2}-1)x\right)^2.$$

More generally, we have
$$B_n(x) = \sum_{k=0}^{n} \binom{n}{k} (e^{1/n}x)^k (1-x)^{n-k} = (e^{1/n}x + (1-x))^n = (1 + (e^{1/n}-1)x)^n.$$

d) It can be shown that $B_n(e^x)$ may be written as $(1 + \frac{x}{n} + \frac{c_n}{n^2})$ where $0 \le c_n \le 1$ and hence $B_n(e^x)$ converges uniformly to e^x.

Solution 11.26

a) Let $\varepsilon > 0$. By the Weierstrass Approximation Theorem, there exists a polynomial $q(x)$ such that $\|q - f\| < \varepsilon/2$. Suppose q has degree r, and
$$q(x) = \sum_{k=0}^{r} a_k x^k$$
where some or all of the coefficients $a_0, a_1, \ldots, a_r$ may be irrational. For each coefficient a_k we find a rational number b_k such that $|b_k - a_k| < \varepsilon/2(r+1)$. Let p be the polynomial given by
$$p(x) = \sum_{k=0}^{r} b_k x^k.$$
Then, for all $x \in [0, 1]$ we have
$$|p(x) - q(x)| = \left|\sum_{k=0}^{r}(b_k - a_k)x^k\right| \le \sum_{k=0}^{r} |b_k - a_k||x|^k \le \sum_{k=0}^{r} \frac{\varepsilon}{2(r+1)} = \frac{\varepsilon}{2}.$$
Then $\|p - q\| \le \varepsilon/2$, so $\|p - f\| \le \|p - q\| + \|q - f\| < \varepsilon$.

b) To show $C[a,b]$ separable we must show it contains a countable dense set. Let
$$A = \{p(x) = \sum_{i=0}^{n} b_i x^i \mid b_i \in \mathbb{Q}, n \in \mathbb{N}\}.$$
Then define
$$A_n = \{p(x) = \sum_{i=0}^{n} b_i x^i \mid b_i \in \mathbb{Q}\},$$

and we have

$$A = \bigcup_{n \in \mathbb{N}} A_n.$$

There is an obvious bijection between A_n and $\mathbb{Q}^{n+1}$ which sends each coefficient of a polynomial in A_n to one coordinate in $\mathbb{Q}^{n+1}$. Thus A_n is countable, and A is countable since it is a countable union of countable sets.

So we must show that A is dense in $C[a,b]$. We follow the same proof as above:

Let $\varepsilon > 0$. By the Weierstrass Approximation Theorem, there exists a polynomial $q(x)$ such that $\|q - f\| < \varepsilon/2$. Suppose q has degree r, and

$$q(x) = \sum_{k=0}^{r} a_k x^k$$

where some or all of the coefficients $a_0, a_1, \ldots, a_r$ may be irrational. Let $c = \max_{0 \le i \le r} \{ \sup_{a \le x \le b} \{|x|^i\} \}$.
Then for each coefficient a_k we find a rational number b_k such that $|b_k - a_k| < \varepsilon/2c(r+1)$.
Let p be the polynomial given by

$$p(x) = \sum_{k=0}^{r} b_k x^k.$$

Then, for all $x \in [0,1]$ we have

$$|p(x) - q(x)| = \left| \sum_{k=0}^{r} (b_k - a_k) x^k \right| \le \sum_{k=0}^{r} |b_k - a_k| |x|^k \le \sum_{k=0}^{r} \frac{|x|^k \varepsilon}{2c(r+1)} \le \sum_{k=0}^{r} \frac{\varepsilon}{2(r+1)} = \frac{\varepsilon}{2}.$$

Then $\|p - q\| \le \varepsilon/2$, so $\|p - f\| \le \|p - q\| + \|q - f\| < \varepsilon$. Thus for any $\varepsilon > 0$ and $f \in C[a,b]$ we can find a $p \in A$ with $\|p - f\| < \varepsilon$, so A is dense in $C[a,b]$.

Solution 11.27

Let $P_n(x) = a_0 + a_1 x + \cdots + a_n x^n$ be a polynomial. Then by the given hypothesis,

$$\int_a^b f(x) P_n(x) dx = a_0 \int_a^b f(x) dx + a_1 \int_a^b x.f(x) dx + \cdots + a_n \int_a^b x^n.f(x) dx = 0.$$

By the Weierstrass Approximation Theorem, we know that any continuous function on $[a,b]$ can be uniformly approximated by polynomials. Hence there exists a sequence $\{P_n(x)\}$ which converges uniformly to $f(x)$ on $[a,b]$. Since $P_n \rightrightarrows f$, then $fP_n \rightrightarrows f^2$. Therefore

$$\lim_{n \to \infty} \int_a^b f(x) P_n(x) dx = \int_a^b \lim_{n \to \infty} f(x) P_n(x) dx = \int_a^b f^2(x) dx = 0.$$

Continuity of $f(x)$ implies that $f = 0$ constantly (see Problem 7.19).

Bibliography

- Abbott, Stephen, *Understanding Analysis.* Undergraduate Text in Mathematics, Springer-Verlag, New York, 2001.

- Amann, H., Escher, J., *Analysis I.* Birkhauser Verlag, Basel, 1998.

- Bartle, Robert G., *The Elements of Real Analysis*, Second Edition. John Wiley and Sons, New York, 1964.

- Beals, Richard, *Analysis: An Introduction.* Cambridge University Press, Cambridge, 2004.

- Bressoud, David, *A Radical Approach to Real Analysis.* The Mathematical Association of America, Washington, DC, 1994.

- Buck, R. C., *Advanced Calculus.* McGraw-Hill, New York, 1965.

- Carothers, N. L., *Real Analysis.* Cambridge University Press, Cambridge, 2000.

- Davidson, K. R., Donding, A.P., *Real Analysis with Real Applications.* Prentice-Hall, Englewood Cliffs, NJ, 2002.

- Dieudonne, Jean, *Foundation of Modern Analysis.* Prentice-Hall, Englewood Cliffs, NJ, 1966.

- Giles, J. R., *Introduction to the Analysis of Metric Spaces.* Cambridge University Press, Cambridge, 1987.

- Hairer, E., Wanner, G., *Analysis by Its History.* Undergraduate Text in Mathematics, Springer-Verlag, New York, 2008.

- Halmos, Paul R., *Naive Set Theory.* Undergraduate Text in Mathematics, Springer-Verlag, New York, 1974.

- Hardy, G., Littlewood, J. E., Polya, G., *Inequalities*, Second Edition. Cambridge Mathematical Library, 1934.

- Janke, Hans N, (Editor) *A History of Analysis.* AMS, LMS, HMATH 24, 2003.

- Marsden, J. E., Hoffman, M. J., *Elementary Classical Analysis*, Second Edition. Freeman and Company, New York, 1993.

- Mattuck, A., *Introduction to Analysis.* Prentice-Hall, Englewood Cliffs, NJ, 1999.

- Pugh, Charles C., *Real Mathematical Analysis.* Undergraduate Text in Mathematics, Springer-Verlag, New York, 2002.

- Rudin, Walter, *Principles of Mathematical Analysis*. International Series in Pure and Applied Mathematics, McGraw-Hill, New York, 1964.

- Simmons, G. F., *Introduction to Topology and Modern Analysis*. International Student Edition. McGraw Hill, New York, 1963.

- Sprecher, David, *Elements of Real Analysis*. Dover Publications, New York, 1970.

- Wade, William R., *Introduction to Analysis*. Prentice-Hall, Englewood Cliffs, NJ, 2000.

Index

Abel's Test, 133
Abel's Theorem, 163, 176
Abel, Niels Henrik, 223
accumulation point of a set, 63, 197
    in a metric space, 197
Al-Khwarizmi, Musa, 159
approximation
    minimax, 230
    uniform, 82
Archimedean Property, 22, 33, 36, 210
Arithmetic and Geometric Means, 25
Arzelà Theorem, 134
Arzelà–Ascoli Theorem, 224, 244

Baire's Category Theorem, 204
Banach Contraction Mapping Theorem, 83, 103
Bernstein Polynomials, 230
Bertrand series, 162, 172, 179
Binomial Theorem, 103
Bolzano–Weierstrass Theorem, 41, 89
boundary of a set, 200
bounded, 22
    above, 23, 42
    below, 23
    function, 63, 70
    sequence, 41
    set, 23
        above, 22
        below, 22
    totally, 198
    uniformly, 229

Cantor function, 93
Cantor set, 24, 202
    compact, 202
    disconnected, 203
    interior, 202
    uncountable, 24
Cantor's diagonalization, 17

Cantor's Intersection Theorem, 201, 203, 212
Cantor–Bendixson Theorem, 205
Cantor–Bendixton Derivative, 204
cardinality, 3, 6
Cartesian product, 3, 204
Cauchy criterion
    for integral, 133
    for series, 173
    for uniform convergence, 226, 237, 244
Cauchy sequence, 42, 81, 181, 184
    in a metric space, 184
    subsequence, 43
Cauchy, Augustin Louis, 63
Cauchy–Schwartz Inequality, 25
Cesaro Average, 44
closed ball
    in a metric space, 185
closed set, 197
    in a metric space, 197
closure of a set, 199–201
compact, 197, 201
    Cantor set, 202
    Cantor-intersection property, 201
    closed subset, 201
    continuous image of, 202
    distance to, 215
    in a metric space, 201
    sequential, 198
    sequentially compact, 201
    subsets of space of continuous functions, 224
complement, 3
    relative, 3
connected, 198, 203
    but not path connected, 204
    continuous image of, 204
continuity, 77, 78
    and periodic, 81, 82
    equicontinuous, 224
    everywhere but nowhere differentiable, 229

Lipschitz, 79
　　　modulus of continuity, 78, 80
　　　nowhere, 80
　　　on a compact set, 202
　　　on a set, 77
　　　on compact sets, 80, 93
　　　pointwise limit, 226
　　　sequential, 79
　　　topological characterization of, 202
　　　uniform, 77, 81
convergence
　　　metric space, 181
　　　not uniform, 225
　　　pointwise, 223
　　　uniform, 225
　　　uniformly, 223, 226
countable, 3, 8, 66, 75, 130, 139, 198, 199
　　　$\mathbb{Q}$ is, 8
　　　$\mathbb{R}$ is not, 8
　　　Cantor set is not, 24
　　　subsets, 7
　　　unions, 8

De Morgan's Laws, 6, 198, 206
Dedekind, Richard, 77
dense, 23, 137
derivative
　　　nth, 100, 101
　　　and tangent line, 97, 98
　　　and uniform convergence, 229
　　　at a point, 97
　　　bounded, 98
　　　does not exist, 107
　　　exists, 99
　　　exists everywhere, 100, 101
　　　Leibnitz formula, 103
　　　nowhere differentiable, 229
　　　on a set, 97
　　　uniformly continuous, 99
　　　uniformly differentiable, 99
diameter of a set, 185
Dini's Theorem, 228
Dirichlet's function, 65, 79, 87
Dirichlet's Rearrangement Theorem
　　　rearrangement, 164
disconnected, 203
　　　Cantor set, 203
discontinuous everywhere, 79

equicontinuous, 224, 228, 229
Euclidean distance, 187
Euler constant, 50
Euler's series, 161, 170
Euler, Leonhard, 41

Fatoo Lemma, 134
Fibonacci numbers, 23
fixed point, 81
function
　　　C^1, 98, 131
　　　C^2, 98, 131
　　　asymptotic to, 67
　　　bijection, 3
　　　bounded, 182
　　　constant, 101
　　　convex, 77, 82
　　　greatest integer, 52, 80
　　　inverse, 215
　　　Lipschitz, 79
　　　monotone, 63, 66
　　　nonconstant, 82
　　　one-to-one, 3, 7, 101
　　　onto, 3
　　　periodic, 65, 78
　　　square-root, 24
　　　step, 82, 130
Fundamental Theorem of Calculus, 128

Hölder inequality, 25
harmonic series, 161
Heine–Borel Theorem, 198, 211, 215
homeomorphism, 202

inf$A$, 22, 24, 32
infinite series, 159
　　　absolute convergence, 159
　　　and termwise differentiation, 224
　　　and termwise integration, 224
　　　Bertrand, 179
　　　conditional convergence, 160
　　　convergence, 159
　　　divergence, 159
　　　Euler, 161
　　　harmonic, 161
　　　partial sum, 159
　　　Raabe–Duhamel's rule, 165
　　　uniform convergence, 227
　　　with positive terms, 162

INDEX

integral
    Bertrand, 133
    improper, 128, 132, 133
    lower, 128, 129, 152
    pointwise limit, 226
    Riemann, 128–130
    Riemann sums, 128
    upper, 128, 129
    Wallis, 45, 59
interchange of limit
    and derivative, 226, 227
    and integral, 226, 227
interior of a set, 200
interior point of a set, 200
Intermediate Value Theorem, 77, 101, 115, 147
    generalized, 204
Interval Intersection Property, 26
irrational, 24, 68, 204

Laplace, Pierre-Simon, 127
Legendre polynomials, 104
Leibnitz formula, 103
limit
    inferior, 42, 45, 46
    is irrational, 57
    of a function, 64
    of a sequence, 41
    of integrals, 131
    of rational numbers, 45
    one-sided, 64
    Riemann sums, 128, 130
    sequential characterization, 65, 92
    superior, 41, 45, 46
limit point of a set, 199, 201
Littlewood, John, 197
Luxemburg Monotone Convergence Theorem, 133, 153, 154

Mean Value Theorem, 97, 108, 112–114, 118
    for integrals, 128, 142, 151
    generalized, 116
measure of noncompactness, 203
metric, 181, 182
    bounded, 183
    discrete, 182, 185
    Euclidean, 182
    Hausdorff, 203
    p-adic, 184
    SNCF, 183
    ultrametric, 184, 190
metric space, 181
    complete, 181, 185
    not complete, 184, 185
    of continuous functions, 182, 183
    ultrametric space, 184
Minkowski Inequality, 25
Monotone Convergence Theorem, 133

natural numbers, 203
    coprime, 25, 36
negation of a statement, 4
neighborhood, 25
nest of intervals, 22
Nested Intersection Property, 26
Newton, Isaac, 97
nowhere dense, 219

open ball
    closure of, 199
    in a metric space, 185, 199
    in Euclidean space, 185

path connected, 198, 204
Picard iteration, 134, 155
power series, 227
    uniform convergence of, 227
power set, 23
proof
    by contradiction, 2, 27, 219
    by induction, 21–23, 37
    by strong induction, 21, 29
    contrapositive, 2

Raabe–Duhamel's rule, 165
rational number, 22, 23, 203
real number, 23, 24
rearrangement, 176
Riemann sum, 128, 144, 145
Riemann–Lebesgue's lemma, 132
Rolle's Theorem, 97, 112, 115

Schröder–Bernstein Theorem, 4, 8
sequence, 41
    bounded, 41, 184
    Cauchy, 42, 43, 49, 219
    convergent, 41–43, 45

decreasing, 42
divergent, 41, 43, 44
in a metric space, 184
increasing, 42
not bounded, 54
ratio, 44
series of functions
    converge pointwise, 223
    converge uniformly, 223
set, 1
    both open and closed, 198
    closed, 198–200
    closed and bounded but not compact, 201
    compact, 201
    countable, 7
    disjoint, 15
    disjoint open, 198
    empty, 7, 25
    equal, 1
    intersection, 2
    open, 198, 200
        in $\mathbb{R}$, 197
    subset, 1
    union, 2

sets
    intersection and union of a family, 3
sphere
    in a metric space, 185
Squeeze Theorem, 51, 52, 56, 66, 112, 142–144
Stirling formula, 45
subsequence, 41, 42, 48, 184
    uniformly convergent, 229
sup$A$, 22, 31

Taylor's Theorem, 98, 143
Tchebycheff polynomials, 135
triangle inequality, 38, 181
    generalized, 189, 191

uniformly continuous extension, 82
unit balls in $\mathbb{R}^2$, 185

Weierstrass Approximation Theorem, 225, 247
Weierstrass M-Test, 224, 227, 238, 241
Weierstrass's example, 229
well-ordering property, 22

Young Inequality, 25

CPSIA information can be obtained at www.ICGtesting.com
Printed in the USA
LVOW05*0732230615

443365LV00002BA/147/P

9 781441 912954